Essential Oils and Fragrances in Cosmetics

This book comprehensively reviews essential oils and fragrances and their role in cosmetics and personal care products. The initial chapters define essential oils, outline their types, and discuss their widespread use in cosmetics, delving into the specific fragrance compounds that contribute to personal care products' sensory and functional attributes. The book addresses critical safety and toxicological assessments, including risk assessment methodologies, toxicological testing requirements, and allergen labeling and management. It emphasizes good manufacturing practices (GMP), providing guidelines and certification processes for product quality and safety. The book provides an overview of key international regulatory bodies and standards, such as International Organization for Standardization (ISO) and International Fragrance Association (IFRA), and discusses efforts toward harmonization and establishing international standards. It presents strategies for global compliance, including case studies of multinational companies that have successfully navigated regulatory challenges. Furthermore, the book examines innovations and regulatory responses, particularly the impact of new technologies like synthetic biology. This book is aimed at professionals in the cosmetic industry, such as product developers, regulatory specialists, and marketing professionals.

Key Features

- Explores the essential oils and fragrances in detail, covering definitions, types, and their roles in cosmetics and personal care products
- Guides safety assessments, GMP, labeling, and advertising regulations
- Reviews key international regulatory bodies like ISO and IFRA and discusses efforts toward standard harmonization
- Includes case studies of multinational companies and offers strategies for global compliance
- Examines the impact of new technologies like synthetic biology and current trends toward sustainability and transparency

Essential Oils and Fragrances in Cosmetics

Global Regulatory Practices and Compliance

Edited by
Ashish Dwivedi
A.B. Pant
Shakti Vinay Shukla
Ruchi Pandey

CRC Press
Taylor & Francis Group
Boca Raton London New York

CRC Press is an imprint of the
Taylor & Francis Group, an **informa** business

Designed cover image: Shutterstock

First edition published 2026
by CRC Press
2385 NW Executive Center Drive, Suite 320, Boca Raton FL 33431

and by CRC Press
4 Park Square, Milton Park, Abingdon, Oxon, OX14 4RN

CRC Press is an imprint of Taylor & Francis Group, LLC

ISBN: 9781032904801 (hbk)
ISBN: 9781032905372 (pbk)
ISBN: 9781003558477 (ebk)

DOI: 10.1201/9781003558477

Typeset in Sabon
by KnowledgeWorks Global Ltd.

Contents

Preface

This book provides a comprehensive exploration of essential oils and fragrances in the cosmetics industry, focusing on their roles in cosmetics and personal care products. Beginning with foundational concepts, Chapter 1 defines essential oils, outlines their types, and discusses their widespread use in cosmetics. It also delves into the specific fragrance compounds that contribute to the sensory and functional attributes of personal care products. In Chapter 2, the book traces the historical journey of essential oils and fragrances, examining their traditional uses and the evolution of the industry. This chapter also highlights current market trends and consumer preferences, setting the stage for understanding the industry's contemporary landscape. Chapter 3 addresses critical safety and toxicological assessments, including risk assessment methodologies, toxicological testing requirements, and allergen labeling and management. These are vital for ensuring consumer safety and regulatory compliance. Chapter 4 focuses on good manufacturing practices (GMP), providing guidelines and certification processes essential for maintaining product quality and safety. It underscores the importance of stringent manufacturing standards in the production of cosmetic products. The book then transitions to regulatory aspects in Chapter 5, detailing mandatory labeling requirements, advertising guidelines, and actions against misleading claims. This chapter is crucial for understanding the legal frameworks that govern product presentation and marketing. Chapter 6 offers an overview of key international regulatory bodies and standards, such as International Organization for Standardization (ISO) and International Fragrance Association (IFRA), and discusses efforts toward harmonization and the establishment of international standards. This global perspective is essential for companies operating in multiple regions. Navigating regulatory differences is the focus of Chapter 7, which presents strategies for global compliance and includes case studies of multinational companies that have successfully managed regulatory challenges. Chapter 8 examines innovations and regulatory responses, particularly the impact of new technologies like synthetic biology. It also explores controversies and public perception issues, including debates over natural versus synthetic ingredients and public health concerns. Trends in regulation and industry practices are covered in Chapter 9, offering predictions for future regulatory changes and highlighting industry trends toward sustainability and transparency. This chapter also emphasizes the role of advocacy and consumer education in shaping regulations and industry practices. Finally, Chapter 10 provides a summary of key points and offers best practices for companies navigating global regulations. This conclusion synthesizes the book's insights, providing a valuable roadmap for industry professionals. Overall, this book serves as a vital resource for understanding the complex regulatory landscape of essential oils and fragrances in the cosmetics industry, offering practical guidance for ensuring compliance and fostering innovation.

About the editors

Ashish Dwivedi graduated from CSIR-Indian Institute of Toxicology Research (IITR), Lucknow, Uttar Pradesh, India, specializing in Photobiology and Phototoxicology. He further honed his expertise through post-doctoral fellowships at the Israel Institute of Technology, Israel; Colorado University, USA; and Banaras Hindu University, Varanasi, Uttar Pradesh, India. Leveraging his outstanding research experience, Dr. Dwivedi returned to his parent research organization, CSIR-IITR, where he currently serves as a Senior Scientist in the System Toxicology & Health Risk Assessment Group. With a decade-long research career, Dr. Dwivedi has established himself as a pioneer in the field of phototoxicity and photosafety, particularly focusing on the photosafety of cosmetics ingredients like fragrances, essential oils, preservatives, and hair dyes. Beyond his laboratory work, he is dedicated to advancing contemporary science education, mentoring postgraduate and PhD students, researchers, and university teachers. Dr. Dwivedi's research has yielded numerous publications in high-impact journals. He is an active member of various scientific societies, including the Society of Toxicology (United States), the American Society for Photobiology (United States), the Indian Photobiology Society (India), and the Society of Toxicology (India), reflecting his significant contributions to the scientific community. He has been awarded a prestigious fellowship from SERB-SIRE for working as a Visiting Professor in Michigan State University, MI.

A.B. Pant, a seasoned toxicologist with over three decades of active research experience, is currently Chief Scientist at CSIR-Indian Institute of Toxicology Research, Lucknow, Uttar Pradesh, India. He also holds the title of Professor at the Academy of Scientific & Innovative Research, an institution of national importance established by the Act of Parliament, Government of India. Renowned in the scientific community, Professor Pant's groundbreaking research focuses on the application of human stem cells and induced pluripotent stem cells (iPSCs) in developmental neurotoxicity. He has established numerous in vitro model systems as alternatives to laboratory animals for biomedical research, founding the discipline of "In-Vitro Toxicology" in India. His dedication extends to educating postgraduate and PhD students, researchers, and university teachers in contemporary science. As a lead Good Laboratory Practice (GLP) inspector for the Government of India, Professor Pant serves as an expert for several regulatory bodies, including the Bureau of Indian Standards (BIS), Central Drugs Standard Control Organization (CDSCO), National Accreditation Board for Testing and Calibration Laboratories (NABL), Food Safety and Standards Authority of India (FSSAI), National GLP Compliance Monitoring Authority (NGCMA), Indian Council of Medical Research (ICMR), Department of Science and Technology (DST), and Pharmacovigilance. Professor Pant's contributions have been recognized with numerous awards, including the Vigyan Ratna Award (2010) by the Council of Science & Technology (CST) UP, the Shakuntala Amir Chand Prize (2007) by the ICMR, the National Bioscience Award (2012) by the Department of Biotechnology (DBT), the Prof. KT Shetty Memorial Oration Award (2017) by the Indian Academy of Neurosciences, the Toxicology Promotion Award (2018) by the National Academy of Sciences, and the STOX/ASAW Gold Medal by the STOX India. He is an elected Fellow of several prestigious organizations, including the Academy of Toxicological Sciences (US), the Society of Toxicology (India), the Indian Academy of Neurosciences, the Academy of Sciences for Animal Welfare (India), and the Academy of Environmental Biology (India). Recognized globally, Professor Pant is listed in the "European and United Kingdom Register of Toxicologists" and has been awarded an honorary DSc degree by HNB Medical Education University in Dehradun, Uttarakhand, India. Professor Pant serves as Chairman and Member of various national and international committees and boards of government institutes and bodies. He is also a member of the editorial board of several esteemed scientific journals, underscoring his prominent role in advancing the field of toxicology.

Shakti Vinay Shukla is a distinguished leader in the Indian aroma industry, currently serving as the Director of the Fragrance & Flavour Development Centre (FFDC) in Kannauj, Uttar Pradesh, India. He earned his BSc (Hons.) in Chemistry in 1988 and MSc in Organic Chemistry in 1990 from Banaras Hindu University. In 1992, he joined FFDC Kannauj as a Scientific Assistant and ascended to the position of Principal Director in 2012. Trained both in India and abroad, he has been instrumental in establishing the Fragrance and Flavour Division at FFDC, developing over 300 fragrances and flavors. He has significantly contributed to skill development, conceiving and conducting various training programs, including the pioneering one-year course in Aroma and its Management and a PG Diploma in Aroma Technology. These programs have been developed in collaboration with FRI University, Dehradun, and Mewar University, Chittorgarh, and included the first NSQF compliant courses approved by the National Skill Development Authority, Government of India. He has also been involved in international training, providing expertise to participants from Rwanda, Egypt, Bhutan, Sri Lanka, Bangladesh, and Senegal. His contributions have been recognized with several awards, including the Young Scientist Award by the CST, UP, Ratna Award, and the Lifetime Achievement Award by the Essential Oil Association of India. He has represented India in various international forums, such as the International Fragranze Show in Italy and consultancy visits to Kyrgyzstan and Rwanda. His affiliations include membership in numerous advisory and executive committees, including the BIS and the CSIR-Aroma and Floriculture Missions. Under his leadership, FFDC Kannauj has achieved self-sustainability, marking a significant milestone in the institution's history. His dedication and expertise continue to drive advancements in the aroma industry both nationally and internationally.

Ruchi Pandey works in the Systems Toxicology & Health Risk Assessment department at CSIR-Indian Institute of Toxicology Research, Lucknow, Uttar Pradesh, India. She earned her PhD in Chemistry from Banaras Hindu University, Varanasi, Uttar Pradesh, India and CSIR-Indian Institute of Toxicology Research, Lucknow, Uttar Pradesh, India. Dr. Pandey is a distinguished researcher and educator specializing in chemistry, environmental chemistry, and toxicology. With over six years of research experience, Dr. Pandey has significantly contributed to the scientific community through innovative experimental techniques and analytical methods. Her work is highly regarded, with numerous peer-reviewed research articles published in esteemed international journals. At the National Cancer Center (NCC) in South Korea, she researched copy number variation in systemic lupus erythematosus. As a Senior Research Scholar and PhD Scholar, Dr. Pandey developed cost-effective methods for heavy metal removal from groundwater and studied bioremediation using agro-waste materials. Dr. Pandey has presented her research at numerous conferences and workshops and has been recognized with several awards, including the Snehlata Banerjee Gold Medal (2014). Her ongoing contributions continue to make a significant impact in both the scientific and educational communities.

Chapter 1

From nature to care

Essential oils and fragrances in cosmetics

Parshant, Heena Choudhary, and Ravindra Shukla

Laboratory of Bioresource Technology, Department of Botany, Indira Gandhi National Tribal University, Amarkantak, India

1.1 INTRODUCTION

Cosmetics have existed for as long as human development and civilization have existed. Every individual attempts to be unique in many aspects of life, both socially and physically. The first records of cosmetic ingredients and their use date back to the Indus Valley civilization between 2500 and 1550 BC; however, the ancient discipline of cosmetology is said to have developed in Egypt and India (Patkar, 2008). The Greek term "kosm tikos" means to have authority over, organize, or expertise in decorating, and it is where the word "cosmetic" originated (Bijauliya et al., 2017). The evolution of cosmetics closely aligns with the progress in chemistry, materials, and packaging, providing valuable insights into the historical and present-day methods of health and beautification (Draelos, 2000). The notion of cosmeceuticals, which integrate the advantages of cosmetics and therapeutics, arose throughout the 20th century, illustrating the enduring connection between nutrition, skin health, and aging (Vollmer et al., 2018). The prolonged and consistent use of cosmetics has resulted in both advantageous and unfavorable consequences, underscoring the importance of meticulously evaluating substances and methods (Chaudhri & Jain, 2009). Chemical-based cosmetics contain detrimental compounds such as parabens, triclosan, benzalkonium chloride, 1,4-dioxane, formaldehyde, and heavy metals. These substances can induce cytotoxicity, genotoxicity, neurotoxicity, mutagenicity, and estrogenicity (Bilal & Iqbal, 2019). Prolonged exposure to cosmetic chemicals through skin absorption can cause abnormal metabolism and physiology, resulting in chronic health problems such as osteomalacia, irritated eyes, and damage to the intestinal mucosa (Gowhar, 2018). Cosmetically, chemicals can cause contact dermatitis, allergic responses, and skin irritation in the short term, while long-term exposure can cause more severe diseases such as carcinomas and endocrine disruption (Samanta, 2021). Due to their diverse advantages, the cosmetic industry increasingly utilizes essential oils (EOs) and their compounds (EOCs) (Kivrak et al., 2017). These natural substances are prized for their fragrant qualities and functional attributes, including antimicrobial, anti-inflammatory, and antioxidant properties. The rising consumer

DOI: 10.1201/9781003558477-1

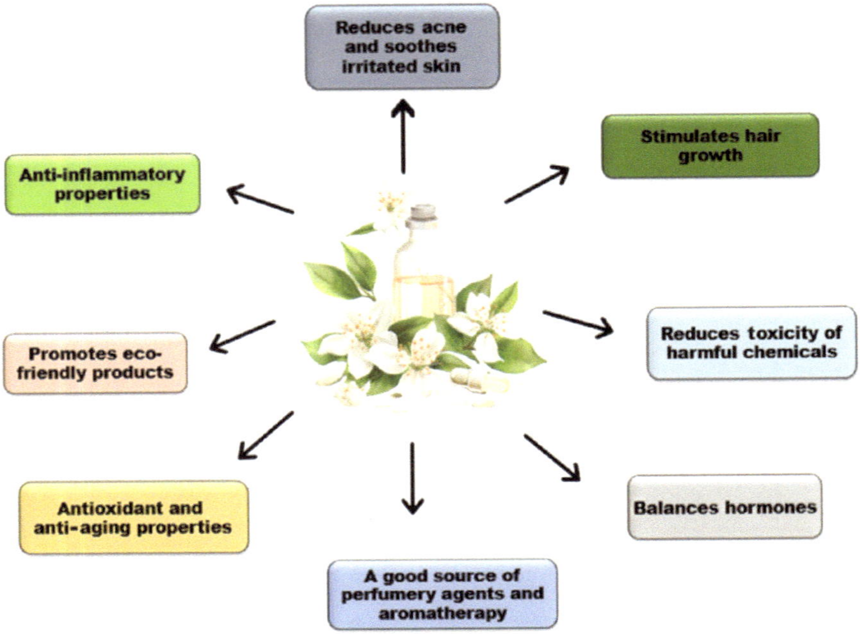

Figure 1.1 Significance of essential oil uses in cosmetics.

preference for natural and eco-friendly products has led to integrate EOs in various cosmetic formulations (Padole et al., 2022). The market for EOs-based cosmetics and cosmeceutical industries is estimated to be worth USD 1.6 billion in 2020, and it is predicted to grow at a rate of 11.6% per year through 2028 (Tripathi et al., 2023). Due to their potent antimicrobial and antifungal properties, EOs (Figure 1.1) are highly effective in combating skin pathogens and addressing conditions such as dandruff, seborrheic dermatitis, and various fungal infections (Happy et al., 2021). EOs with potent anti-inflammatory and antioxidant effects, such as oregano, lavender, and eucalyptus, aid in lowering oxidative stress and inflammation in the epidermis and encourage healthier skin (Avola et al., 2020). Various research has indicated that certain EOs may potentially prevent and treat various forms of skin cancer, including melanoma and non-melanoma types. These oils exhibit chemopreventive and therapeutic properties that could be beneficial in combating these skin conditions (Pavithra et al., 2019).

EOs and their compounds play a significant role in developing chemical-free cosmetic products (Figure 1.2). However, their application is often limited due to their volatility and instability. Nanotechnological techniques such as nanoemulsion and nanoencapsulation protect the degeneration of bioactive components of EOs and ensure the effective release of application in cosmeceuticals (Jamir et al., 2024; Medeleanu et al., 2023). This chapter investigates

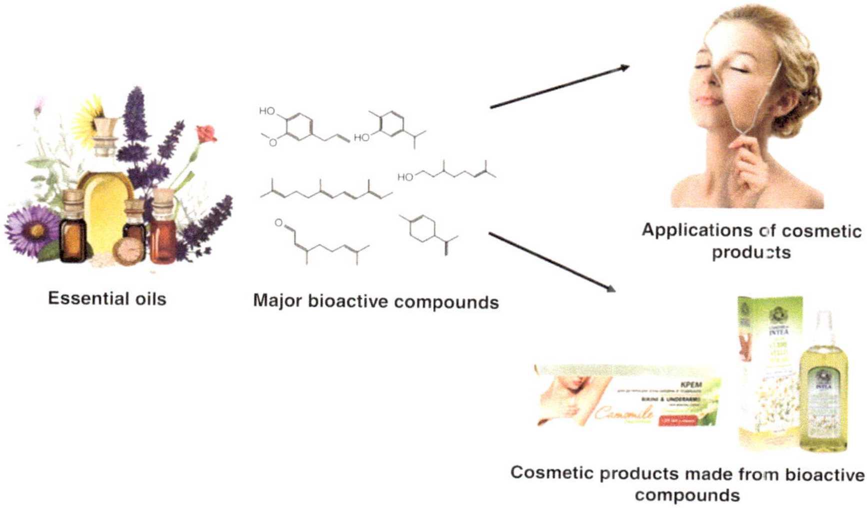

Figure 1.2 Bioactive compounds of EOs and their use in cosmetics.

the contemporary viewpoint on many applications of EOs in the cosmetic sector and each of its constituent parts. It covers a few essential points about the possible safety measures needed when using EOs and EOCs to manufacture toiletries and cosmetics.

1.2 ESSENTIAL OILS AND EXTRACTION METHODS

EOs are sweet and spicy-scented volatile liquids. These are fragrant, oily substances derived primarily from various plant components such as aerial parts. The volume of oil ultimately constitutes 90–95% of the total oil weight. This includes monoterpenes, sesquiterpenes, and their oxygenated derivatives, including aliphatic alcohols, esters, and aldehydes, which may also be present in this category (Tongnuanchan & Benjakul, 2014). Various methods have been employed to extract EOs, enhancing their bioactive properties and therapeutic effects (Figure 1.3). The most effective techniques utilized in these extraction processes are hydro distillation, steam distillation, microwave-assisted hydro distillation, and solvent-free microwave extraction (Herman et al., 2019) (Figure 1.3). The method selected can impact the extraction process's effectiveness, safety, and environmental impact (Malik, 2019).

1.2.1 Hydro distillation technique

In this technique, the plant material is dipped into the bulb, containing sufficient water, and boiled at a specific temperature (60°–100°C). Due to the

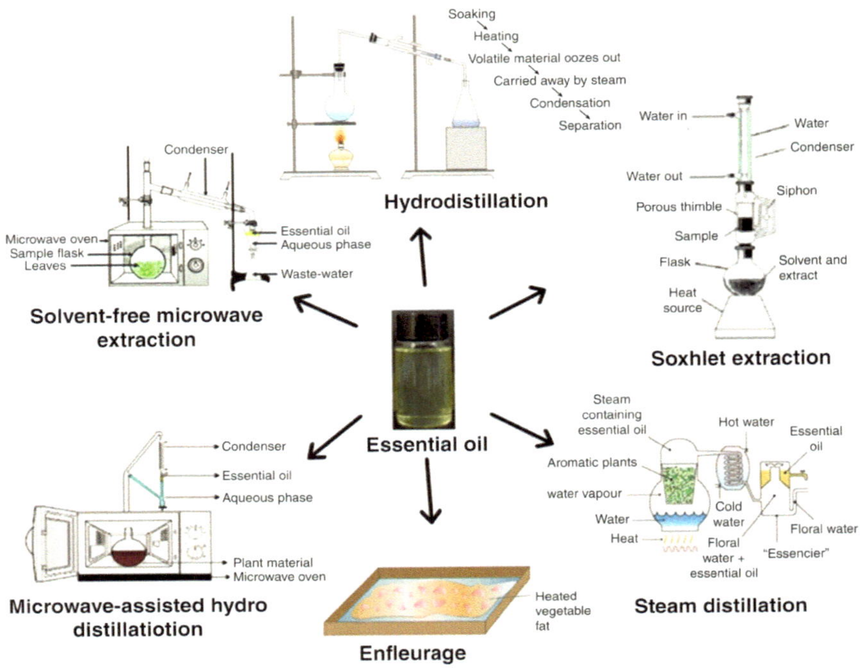

Figure 1.3 Essential oils extraction methods.

effect of the steam, EOs are released from the oil glands of plant material, and condensation occurs. Due to this reaction, the water and oil form two different layers. Finally, EO is extracted through a knob. The primary benefit of this method is that it reduces the setup and running costs, and steam penetrates deeply into the plant material (Dilworth et al., 2024; El Asbahani et al., 2015).

1.2.2 Soxhlet extraction technique

The Soxhlet extractor was invented in the 18th century. The plant powder sample is placed into the thimble and inserted into the Soxhlet extractor. The solvent is added to the plant material, and EO can be isolated after 3–4 hours. The amount of ginger sample was utilized following the statistical design outlined in the Box-Behnken standard design (Belwal et al., 2018; Mohammed et al., 2022).

1.2.3 Stem distillation technique

Steam distillation is a popular technology in the EO business. Steam is produced and provided to the plant material's vessel or apparatus at a specified pressure during steam distillation. Subsequently, as the steam passes through the plant components, the materials' pores allow the volatile molecules to escape. In steam distillation, temperature may be adjusted to manage the organic compounds' breakdown until it falls below the components' decomposition temperature (Mohammad Azmin et al., 2016).

1.2.4 Enfleurage

The enfleurage technique has yet to be applied, although it is a traditional technique used to extract EOs from floral petals. In this method, any vegan fat (Karite fat) or animal fat (Lard from pork) is heated without reaching its melting point. Dried floral petals are added at 40°C for 20 minutes. Cold fat extraction is filtered with 50 ml of ethanol (95%) at 60°C to isolate EOs. Usually, soaps are made with this type of EOs (Ibrahim et al., 2021; Sareriya et al., 2023).

1.2.5 Microwave-assisted hydro distillation technique

Microwave-assisted hydro distillation technique is an advanced technique used to extract EOs. In the development of this application, microwave dielectric heating has been used to achieve effective and targeted heating while reducing thermal gradients. This approach enables a quicker, eco-friendly, and more efficient extraction process, resulting in improved purity and reduced degradation of volatile compounds (Liu et al., 2012; Tran et al., 2019). The basic working principle is dielectric heating of microwave. The composition of EO is the same as that of hydro distillation. Plant materials undergo a quick and gentle breakage of oil glands, resulting in sufficient extraction. The conversion of microwave energy into heat energy in water occurs due to its high dielectric properties, leading to rupturing. This heat is then transferred to plant materials, distributing it in a way that releases the EOs without causing damage or rupture to the EO glands (Desai et al., 2010; Kusuma & Mahfud, 2017).

1.2.6 Solvent-free microwave extraction technique

Solvent-free microwave extraction technique, in particular, is based on integrating dry distillation and microwave thermal energy at atmospheric temperature without adding any water or solvent. The distillate then continually condenses via a cooling mechanism outside the microwave chamber. Condensed water was refluxed to the extraction vessel to provide consistent temperature and humidity levels during extraction. At 100°C, the extraction process was maintained until no more vital oil could be extracted (Lucchesi et al., 2004; Sethunga et al., 2022).

1.3 CHEMISTRY AND BIOACTIVE CONSTITUENTS OF ESSENTIAL OILS

EOs play a valuable and significant role in the ecology and physiology of aromatic plants. They are responsible for (a) attracting pollinators, (b) creating a natural weapon against pathogens, (c) providing an excellent opponent against herbivores, and (d) cell signaling (Holopainen, 2004). Additionally, EO offers multiple health benefits (Wagner & Elmadfa, 2003). The volatile organic compounds that compose fragrant and aromatic plants come from three chemical categories: isoprenoids, fatty acid derivatives, and phenolic compounds (Caissard et al., 2004; Sangwan et al., 2001). The number of isoprenoids, terpenes, or terpenoids doubles every ten years. Over 55,000 have been identified, and some have been assigned different physiological functions (Langenheim, 1994).

1.3.1 Biosynthesis of essential oils

Volatile compounds are formed in aromatic plants via five basic steps: (1) precursor synthesis, (2) primary pathways, (3) formation of basic terpene, (4) modification and diversification, and (5) storage and secretion (Figure 1.4).

During the biosynthesis of EOs, precursor synthesis from primary metabolites is branched into the mevalonate pathway (MVA) and the methylerythritol phosphate pathway (MEP). The formation of isoprenoid units starts with isopentenyl diphosphate (IPP) and dimethylallyl diphosphate (DMAPP). Their condensation synthesizes terpenoid backbones (i.e. monoterpenes, sesquiterpenes, and diterpenes). Modification reactions (oxidation, reduction, hydroxylation, cyclization, and glycosylation) further alter the structure of terpenoid backbones. Finally, volatile and bioactive compounds are formed and stored in any parts of the aromatic plants (leaf, flower, trichome, root, and rhizome) (Chappell, 1995; Rehman et al., 2016).

1.3.2 Volatile components of essential oils

EOs have an intense odor and can include up to 100 low-molecular weight compounds. According to a chemical perspective, they fall into many groups, the primary one being that of oxygenated substances (alcohols, oxides, aldehydes, phenols, ketones, acids, and esters) and terpenes (mono-, sesqui-, and di-terpenes). With a distribution of more than 90%, monoterpenes are among the most typical terpenes. They also have a unique structural variation (Price et al., 2021). Numerous distinct scents and flavors, such as fruity ((E)-nerolidol), floral (Linalool), citrusy (Limonene), herbal (-selinene), and so forth, may be produced using alcohols, aldehydes, and ketones. Aromatic oils include thousands of terpene-family compounds that have been discovered. These substances include phenols (thymol), ketones (menthone, p-vetivone),

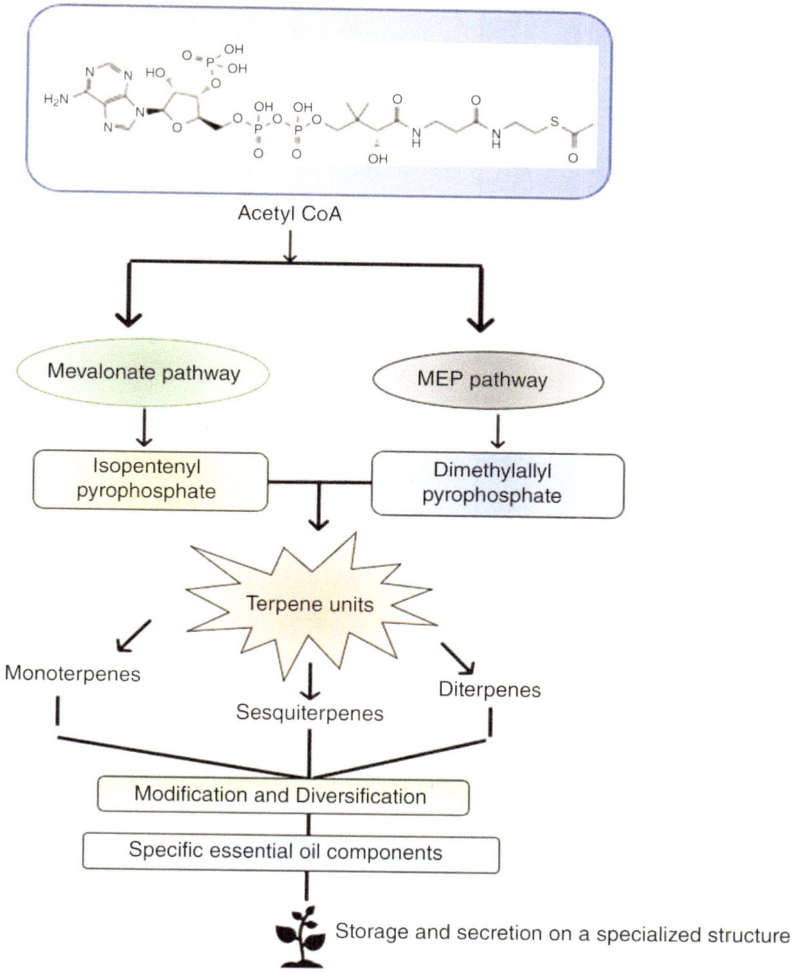

Figure 1.4 Biosynthesis of essential oils in aromatic plants.

esters (-tepinyl acetate, cedryl acetate), and functionalized derivatives of alcohols (geraniol, -bisabolol) (Sharma et al., 2023) (Figure 1.5).

1.4 APPLICATIONS OF ESSENTIAL OILS AND THEIR FRAGRANCE COMPOUNDS IN COSMETIC INDUSTRIES

Volatile chemicals are necessary in cosmetics because they affect product stability, effectiveness, and sensory aspects. Their uses improve the user experience and include active chemicals, scents, and preservatives. EOs and EOCs

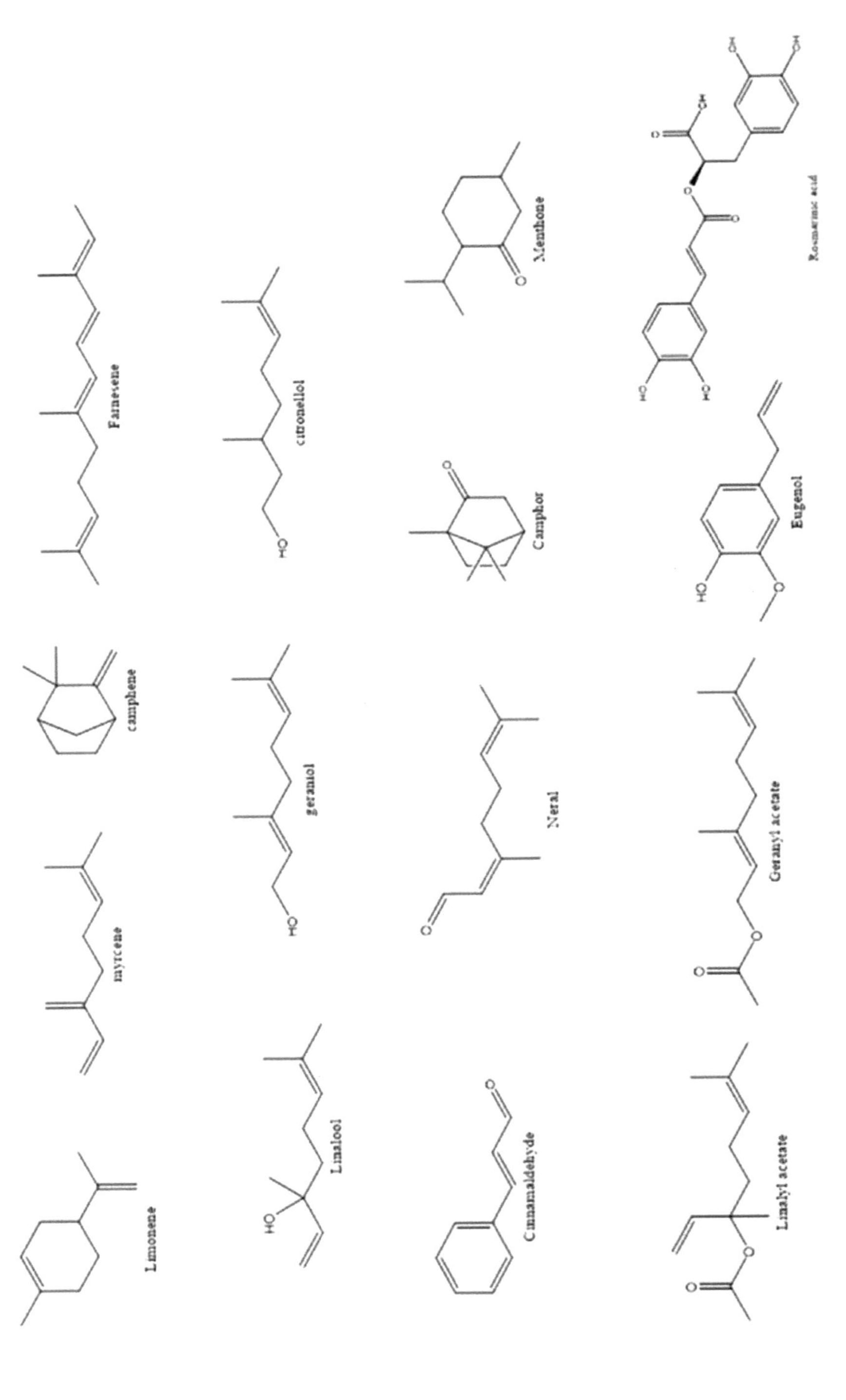

Figure 1.5 Essential oils' bioactive volatile compounds.

have recently gained much popularity as skin care product components. The use of these molecules in skin care products is becoming more popular as a way to use some of their biological qualities, such as antioxidant, antibacterial, and anti-inflammatory effects, which help keep skin looking young and healthy and protect it from environmental harm (Aburjai & Natsheh, 2003).

1.4.1 Linalool

Aromatic monoterpene alcohol linalool (2,6-dimethyl-2,7-octadien-6-ol) is a common constituent of EOs. It is frequently used as an addition to food, cosmetics, household cleansers, and fragrances. It is a colorless to extremely pale-yellow liquid with an anticipated vapor pressure of ~0.05 mm Hg at 20°C, flash point of 71°C, boiling temperature of 198°C, refractive index of 1.460–1.463 (20°C), and specific gravity of 0.860–0.864 (20°C) (Letizia et al., 2003). Linalool is mainly found in lavender (*Lavandula angustifolia*): 20%–45%, coriander (*Corianderum sativum*): 62.2%–76.7%, and mint *(Mentha piperita)*: 51.8% (da Silva Ramos et al., 2017; Khatri et al., 2023; Satyal & Setzer, 2020). Linalool offers health benefits such as antimicrobial, soothing fragrance, and anti-inflammatory properties. An experiment of 9 years was conducted in Japan to assess the tolerance of lavender oil having linalool. Batista et al. (2010) checked the linalool's anti-inflammatory effects inducing chronic inflammatory conditions in animal models. They concluded that it also lowered the discomfort brought on by inflammatory illnesses, indicating that it may have broader uses in treating long-term inflammatory disorders. Linalool not only reduces inflammation but also lowers the pain. Linalool is an excellent source of antimicrobial agents, and it can be used in anti-acne creams, serums, and peel masks. De Oliveira Lima et al. (2017) showed that linalool, which is often used to treat dermatophytosis, had a beneficial impact on a strain of *Trichophyton rubrum* that was resistant to fluconazole (an antifungal drug) (Figure 1.6).

1.4.2 Citronellol

Citronellol is an acyclic monoterpene alcohol, an active constituent of EOs. It is a colorless liquid that smells like roses with a boiling point of 200°C, a vapor pressure of ~ 0.009, a refractive index of 1.4543 at 20°C, and a density of 0.8560 g/cm^3 at 20°C (FEMA, 1994; Lide, 2004). Commercially, citronellol is used to manufacture perfumes, insecticide repellents, candles, toilet soaps, and decorative cosmetics (Lapczynski et al., 2008). It is mainly found in rose (*Rosa damascene*): 36.69%, citronella (*Cymbopogon nardus*): 5.8%, and geranium (*Pelargonium graveolens*): 22.4%–42.4% (Dobreva & Nedeltcheva-Antonova, 2023; Nakahara et al., 2013; Wang et al., 2014). Because of its pleasant aroma, citronellol adds a revitalizing element and more richness to perfumes. This characteristic is crucial for producing long-lasting and well-balanced scents. Citronellol has excellent antioxidant effects that help to protect the skin from UV radiation and other harmful environmental effects (Kaur

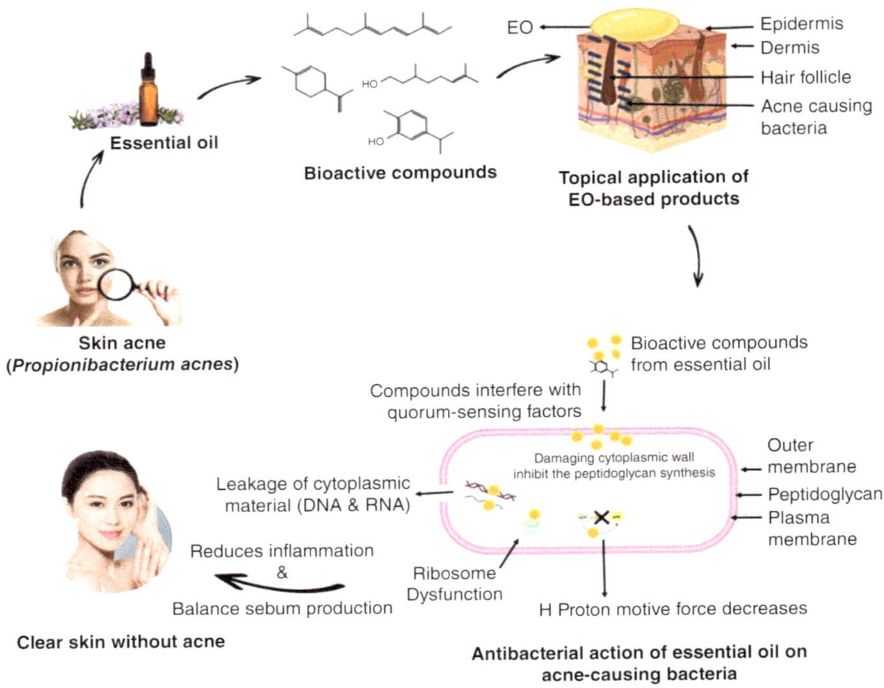

Figure 1.6 Antimicrobial activity of bioactive compounds in essential oils.

et al., 2021a). A study that was published in the *Flavour and Fragrance Journal* states that citronellol's pleasant floral odor that is reminiscent of roses, making it a valuable component in floral perfumes, is the reason it is so commonly employed in the fragrance business (Bauer et al., 2008). Moreover, citronellol is one of the 26 scent allergens that, according to an International Fragrance Association (IFRA) study, must be declared on labels when present in significant concentrations in cosmetic products (Nardelli et al., 2011).

1.4.3 Alpha-bisabolol

Bisabolol or alpha-bisabolol, also called alpha 4-dimethyl-alpha-(4-methyl-3-pentenyl)-3-cyclohexene-1-methanol, is a naturally occurring mono sesquiterpene. It is a colorless liquid with a boiling point of 153°C and a comparatively low density of 0.93g/cm^3. Bisabolol is lipophilic in nature and exhibits a tendency toward oxidation (Perbellini et al., 2004). It is used in aromatherapy products, after-shave creams, body lotions, and deodorants (Gomes-Carneiro et al., 2005). It is mainly found in sandalwood (*Santalum album*): 12.53%, chamomile (*Matricaria recutita*): 52%–60% (Fejer & Salamon, 2016; Horton, 2024). The darkening of a skin region caused by an increase in melanin is known as hyperpigmentation. The primary causes of hyperpigmentation are inflammatory skin conditions, allergic contact dermatitis, and irritating contact dermatitis. It is shown that the cAMP response element (CRE) contributes to the synthesis of alpha-melanocyte-stimulating hormone (αMSH). The depigmentation impact of alpha-bisabolol was investigated using two distinct assays: the CRE luciferase reporter assay and the melanin test (Kim et al., 2008). The findings showed that alpha-bisabolol prevented alpha-MSH-induced CRE activation. Likewise, the substance decreased the amount of melanin produced by alpha-MSH. The enzyme tyrosinase is responsible for catalyzing the synthesis of melanin. Research has shown that αMSH activates the gene expression of microphthalmia-associated transcription factor (MITF), which in turn causes the production of tyrosinase genes. It was found that alpha-bisabolol inhibited the expression of the MITF and tyrosinase genes, suggesting that it inhibits melanogenesis by lowering intracellular cAMP levels. This investigation was carried out to ascertain the effect of alpha-bisabolol on the alpha-MSH-induced expression of MITF and tyrosinase genes. de et al. (2024) investigated that alpha-bisabolol has anti-depressant effects. When compared to the reference medication fluoxetine, the depressed and anxious behavior of treated mice was successfully suppressed following alpha-bisabolol administration (Figure 1.7).

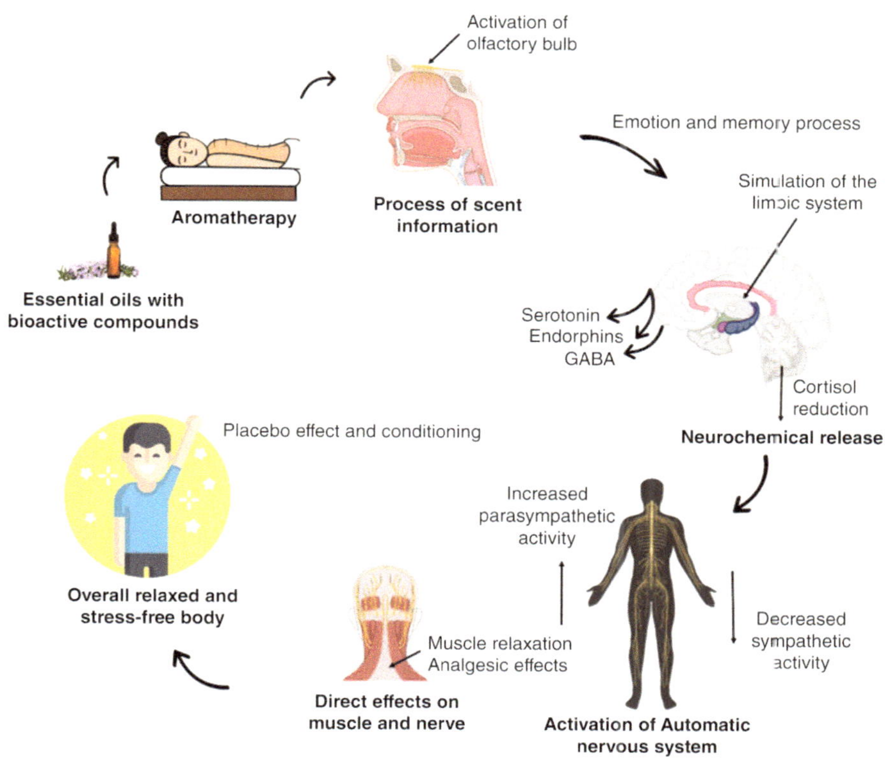

Figure 1.7 Effects of volatile compounds against stress.

1.4.4 Eugenol

Eugenol is an allyl chain-substituted phenylpropene, also known as guaiacol. Eugenol belongs to the group of chemical compounds known as phenylpropanoids. Eugenol is a volatile, colorless yellowish liquid that is a mildly acidic phenolic chemical. At 25°C, its solubility in water is modest, with 2,460 mg l^{-1}, a melting point of −7.5°C, a boiling point of 254°C, and a density of 1.06 g/cm^3 (Yuwono et al., 2002). It is mainly found in clove buds (*Syzygium aromaticum*): 70%–80%, bay leaf (*Laurus nobilis*): 19.4%–21.2%, and cinnamon leaf (*Cinnamomum verum*): 81.7% (Chakraborty et al., 2015; Cortés-Rojas et al., 2014; Peris & Blázquez, 2015). Eugenol makes bath products, antifungal creams, insecticidal formulations, herbal shampoos, and lotions, and it is a good source of antioxidants (Figure 1.8).

Makuch et al. (2020) formulated the cosmetic formulation, examining antioxidant activity and skin permeation ability. Franz diffusion chambers were used in their experiment to measure how well these chemicals penetrate pig skin. The results indicate that eugenol has intense skin penetration and antioxidant activity, suggesting possible uses in cosmetic formulations to counteract oxidative stress and improve therapeutic benefits. The antifungal capability of eugenol and EOs of cinnamon, clove, bay, and pimento, was

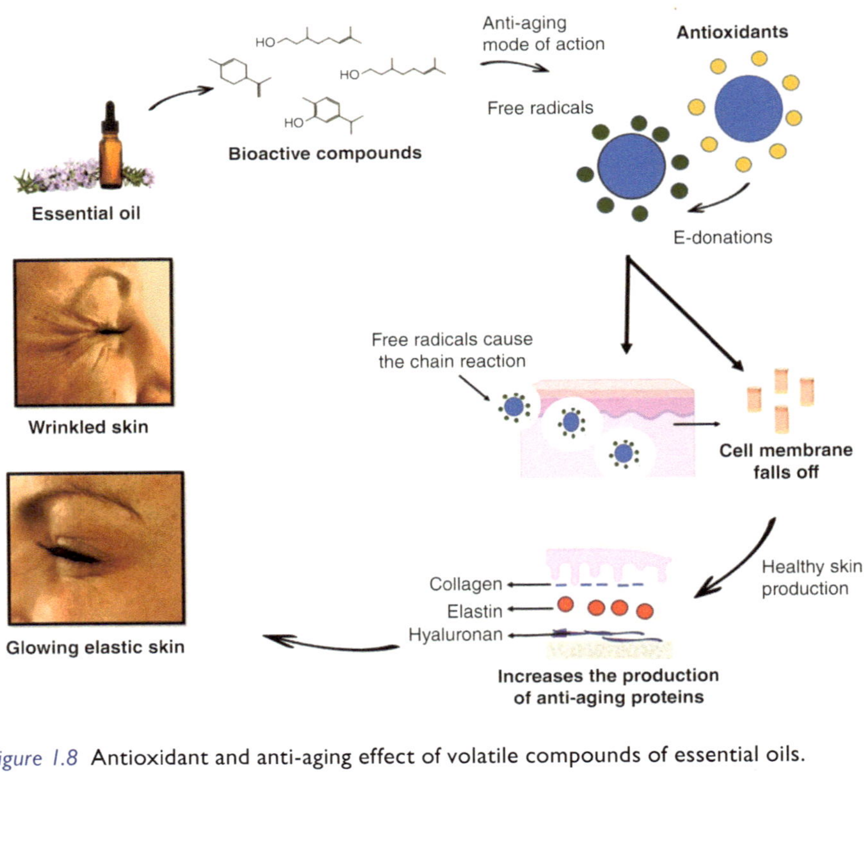

Figure 1.8 Antioxidant and anti-aging effect of volatile compounds of essential oils.

investigated in research against 38 clinically isolated strains of *Candida albicans*. The findings supported the hypothesis that there was a significant correlation between the analyzed EOs' antifungal properties and their eugenol content (Schmidt et al., 2007).

1.4.5 Menthol

Menthol is a simple and natural cyclic monoterpene alcohol. It is commonly called camphor mint. Menthone and isomenthone give a fresh, cooling, and minty taste. Menthol has a flaky appearance and is colorless in nature. At ambient temperature, menthol has a solid density of 0.890 kg/dm^3 (25°C) and a melting point of 41–44°C, depending on its purity. Menthol is readily soluble in alcohol, diethyl ether, or chloroform but only partially soluble in water (435.5 mg/L at 25°C) (Kamatou et al., 2013). It is mainly found in Canadian mint *(Mentha canadensis)*: 60%–75%, and spearmint (*Mentha spicata*): 11.2% (Shiwakoti et al., 2015; Snoussi et al., 2015). Menthol is widely used in cosmetic companies that produce face wash, face masks, lip balms, and hair oil formulations. In addition, menthol is used to make toothpaste, mouthwash, cough syrups, room fresheners, gastrointestinal disorder medicines, antiseptic lotions, and other dermatological formulations (Patel et al., 2007). Applications of menthol, due to its antioxidants and antiseptic properties, are well known. Rozza et al. (2021) investigated how menthol and other natural substances affect wound-healing processes in rats and mice. The results showed that menthol, at a concentration of 0.5%, significantly improves wound closure, lowers inflammation, and stimulates antioxidant activity. Menthol also promotes cell proliferation and decreases inflammatory cytokines for speed healing.

1.4.6 Vanillin

Vanillin is the naturally occurring phenolic aldehyde in the vanilla plant (*Vanilla plantifolia*) with a chemical name 4-hydroxy-3-methoxy benzaldehyde. It is white or pale yellow with a pleasant and aromatic odor. The boiling point of vanillin is 285°C, and the melting point is 83°C. Its flash point is 153°C, and its density is 1.056g/cm^3 (Anand et al., 2019). It is mainly found in Vanilla pods from 1% to 30%. The content of vanillin varies due to the different varieties and extraction methods (Wang et al., 2022). Generally, vanillin is a flavoring agent in the food industry and culinary practices. Besides this, its pleasant aroma attracts the cosmetic industry for skin care products such as body wax formulations, soap bombs, perfumes, lipsticks, and other dermatological products for clear and glowing skin. Cheng et al. (2017) experimented the effects of vanillin on imiquimod-induced psoriasis and skin inflammation in BALB/c mice. They found that vanillin dramatically lowers immune cell infiltration, inflammatory cytokine levels, and epidermal thickness. Vanillin plays a critical role in the pathophysiology of psoriasis by altering the IL-23/IL-17s, as demonstrated by their research using histological examination and gene expression profiling. The results thus indicated that vanillin control inflammatory reactions and cytokine levels, which may make it a potential therapeutic agent with low toxicity for treating psoriasis.

1.4.7 Citral

Citral is an acyclic monoterpene aldehyde made up of two isoprene units. Citral is a pale-yellow color liquid with a lemon aroma. The boiling point of citral is 230°C, density is 0.888 g/cm^3 at 25°C, and flash point is around 72°C (Linstorm, 1998). Citral is commonly found in lemongrass (*Cymbopogon citratus*), in a mixture of terpenoid and geranial: 70%–80%, and lemon myrtle (*Backhousia citriodora*): 90%–96% (Rauber et al., 2005; Southwell, 2021). Citral has several properties, such as antimicrobial and anti-inflammatory properties. In addition, it is vastly used in the food and cosmetic industries. It is used to make body lotions, creams, perfumes, waxing strips, and sun-induced tan products (Golda, 2022). Kim et al. (2018) conducted a clinical trial of citral-containing EO on Korean women against acne skin. The experiment was conducted on 20 women who had acne skin. Unlike the control groups, which showed little changes, the experimental groups utilizing citral-containing EO showed substantial improvements in several indices, including exceptional pores, blackheads, and sebum production. Gaspar et al. (2022) designed a biofilm with citral-dominated lemongrass essential oil (LEO) for possible cosmetics and food preservation usage. These cosmetic formulations have been found to be excellent sources of anti-aging cosmeceuticals because of their remarkable antibacterial and antioxidant capabilities.

1.4.8 Eucalyptol

Eucalyptol or 1,8-cineole is a monocyclicterpene ether (oxide). It is a colorless liquid with a camphor-like aroma. Its boiling point is 176–177°C, and its melting point is 2.9°C. The density of eucalyptol is 0.9225 g/cm^3. Eucalyptol is insoluble in water and soluble in organic solvents. It is commonly found in different Eucalyptus species (i.e. *Eucalyptus globules*, *Eucalyptus polybractea*): 90%–95% and rosemary (*Rosmarinus officinalis*): 17%–24%. Eucalyptol's ratio depends upon the cultivars' varieties (Bhowal & Gopal, 2015; Micić et al., 2021). It has various significant effects against acne, promotes hair growth and wound healing, and is a good source of antioxidants. According to reports, the usual amounts of eucalyptol in cosmetic items are 1.6% in perfume, 0.4% in soap, 0.1% in creams and lotions, and 0.04% in detergents. Rehman et al. (2022) developed and assessed the eucalyptol-based nano gel and examined the wound healing activity. During in vivo investigation, a notable improvement in the wound was noticed. Uronnachi et al.

(2022) formulated an oleogel for hair growth enhancement, containing cedar wood and rosemary essential oils powered with eucalyptol. After six weeks of the treatment, 10% rosemary oil had comparable results to the positive control (2% minoxidil).

1.4.9 Limonene

Limonene is an aliphatic hydrocarbon classified as a cyclic monoterpene, and it exists in two active forms. l-Limonene has pine-like, and d-limonene has a pleasant orange-like fragrance and is a colorless liquid in appearance (Nikfar & Behboudi, 2014). Limonene is mainly found in different citrus peels, and its concentration varies from species to species. Sweet orange (*Citrus sinensis*), bitter orange (*Citrus aurantium*), and lemon (*Citrus limon*) have 95%, 67%–90%, and 67%–69% limonene, respectively (Bourgou et al., 2012; Sikdar et al., 2016). The boiling point of limonene is 177.6°C, the melting point is −74°C, and the flash point is 48°C. Its density is 0.8411 g/cm^3 at 20°C. Limonene offers a variety of health benefits due to its anti-inflammatory properties, antimicrobial properties, and antidepressant activities. Limonene is slightly soluble in water, propylene, and ethanol (Haynes, 2014; Massaldi & King, 1973). Limonene is used in aromatherapy products, massage oil, household cleaning agents, detergents, and food flavor additives in sweet desserts (Dornic et al., 2018; Rubiolo et al., 2010). Miller et al. (2012) assessed the topical application of breast massage oil containing limonene. During the animal disposition test, mice received the oral or topical dosage of limonene in orange EO every day for 4 weeks. Whereas, in the human tolerability study, healthy women were hired for safety evaluation and advised to massage their breasts daily for 4 weeks. After the last dosage, mammary pads and plasma were collected, and the levels of limonene were measured with no discernible toxicity symptoms. Kumar et al. (2022) examined the dermo-protective activity of limonene. Their study evaluated how citrus fruits' natural ingredient, limonene, protects human skin keratinocytes (HaCaT cells) against UVB-induced photodamage and oxidative stress. It shows that limonene increases the production of endogenous antioxidants (i.e. HO-1 and NQO-1), improves cell survival, lowers the creation of reactive oxygen species (ROS), and activates the Nrf2-dependent antioxidant defense mechanism. Furthermore, limonene helps preserve the integrity of tight junction proteins. It suppresses the release of α-melanocyte-stimulating hormone (α-MSH), indicating that it may be used as a hepatoprotective drug for those often exposed to UV light.

1.4.10 Camphor

Camphor is a naturally occurring terpenoid and a cyclic ketone. It is white and waxy in appearance and has a strong aromatic odor. The boiling point of camphor is 204°C. It melts at 180°C, and the vapor density is 5.2. It is soluble in 0.125 g/100 ml of water at 25°C and organic solvents, sometimes used as a carrier in oils (Wickstrom, 1988). It is mainly found in the camphor plant (*Cinnamomum camphora*): 51.3%. Camphor is widely used in cosmetic formulations as an ingredient in lotions, antidandruff formulations, and antimicrobial ointments (Babaei et al., 2021; Guo et al., 2016). Camphor is an excellent source of antimicrobial agents against skin diseases. Karaca et al. (2021) investigated the potential therapeutic interactions between *Staphylococcus aureus*, camphor, and *Lavandula latifolia* EO. Synergistic effects were seen due to the high activity of the combination of EOs and (+)-camphor (concentrations 0.31 + 1.25 mg/mL, respectively). Bekka-Hadji et al. (2022) examined the activity of *Artemisia herba-alba* EO, having 32% camphor to imipenem-resistant *Acinetobactor baumannii* in light of the rising prevalence of antibiotic resistance. Furthermore, they demonstrated that EO and ticarcillin-clavulanate, cefotaxime, imipenem, and nalidixic acid work synergistically to combat *A. baumannii*. Table 1.1 summarizes the plant EOs and their major bioactive compounds with cosmeceutical properties.

Table 1.1 List of aromatic plants, their essential oils, and the primary bioactive compounds with promising properties for skin care and personal care

S. no.	*Botanical name*	*Common name*	*Major bioactive compounds*	*Cosmeceutical properties*	*References*
1.	*Melaleuca alternifolia*	Tea tree	Terpinen-4-ol, γ-terpinene, α-pinene, eucalyptol, and linalool	Anti-inflammatory, antiacne, antifungal	Hammer et al. (1999)
2.	*Cedrus deodara*	Cedar tree	β-himachalene, α-himachalene, and γ-himachalene	Anti-hair loss, antibacterial, antioxidant, and anti-cancerous	Chaudhary et al. (2009); Chauiyakh et al. (2023)
3.	*Citrus bergamia*	Bergamot	Limonene, linalool, and linalyl acetate	Aromatherapy, anti-proliferative, and antimicrobial activity	Navarra et al. (2015)
4.	*Vetiveria ziznaioides*	Vetiver	Khusimol and vetiselinenol	Air-freshener, antioxidant, and antibacterial	Chahal et al. (2015)
5.	*Myrtus communis*	Sweet myrtle	A-pinene, 1,8-cineole, and linalool	Antidermatophytes, inhalators and astringent	Bouzabata et al. (2015)
6.	*Zingiber officinalis*	Common ginger	α-zingiberene, β-sesquiphyllandrene, and α-curcumene	Antiphotoaging, antimicrobial, and immunomodulator	Mahboubi et al. (2019)
7.	*Thymus vulgaris*	Thyme	Thymol, γ-terpinene, and P-cymene	Antitumor, antioxidant, and antimicrobial	Nikolić et al. (2014)
8.	*Cymbopogon martini*	Palma rosa	Geraniol and geranyl acetate	Wound healing, antiyeast, and anti-clastogenic	Ganjewala (2009)

(*Continued*)

Table 1.1 (Continued) List of aromatic plants, their essential oils, and the primary bioactive compounds with promising properties for skin care and personal care

S. no.	*Botanical name*	*Common name*	*Major bioactive compounds*	*Cosmeceutical properties*	*References*
9.	*Origanum vulgare*	Oregano	Carvacrol and Thymol	Anticancer, tissue remodeling, and antioplastic	Han & Parker (2017)
10.	*Ocimum basilicum*	Basil	Linalool and Eugenol	Anticonvulsant, antimicrobial, and cooling effects	Poonkodi (2016)
11.	*Cupressus sempervirens*	Italian cypress	α-pinene and δ-3 carene	Antiseptic, aromatherapeutic, and antioxidant	Patgar et al. (2021)
12.	*Styrax benzoin*	Gum Benjamin	Benzyl benzoate and Cinnamic acid	Antidermatophytic activity	Kholibrina & Aswandi (2021)
13.	*Foeniculum vulgare*	Fennel	Trans-anethole	Antioxidant, antimicrobial, and skin soothing agent	Rather et al. (2016)
14.	*Pogostemon cablin*	Patchouli	Patchoulol and α-bulnesene	Free radical scavengers, blood coagulants, and antimicrobial	Swamy & Sinniah (2015)
15.	*Pelargoinum graveolens*	Geranium	Citronellol and geraniol	Antioxidant, antimicrobial	Kačániová et al. (2023)
16.	*Thuja occidantalis*	White cedar	Thujone	Anti-cancerous, antimicrobial, and antioxidant	Jasuja et al. (2015)
17.	*Cananga odorata*	Ylang-ylang	Germacrene-D, linalool, and benzyl acetate	Anesthetic, psychic energizer	Borgonetti et al. (2022)
18.	*Hyssopus officinalis*	Hyssop	Pinocamphol, limonene, and β-pinene	Myorelaxation activity, antifungal, and antibacterial	Sharifi-Rad et al. (2022)
19.	*Anethum graveolens*	Dill	Carvone, limonene, and α-phellandrene	Antioxidant and antimicrobial	Kaur et al. (2021b)
20.	*Rosa damascene*	Damask rose	Citronellol and Geraniol	Anticonvulsant and antimicrobial	Akram et al. (2020)
21.	*Helichrysum itlalicum*	Italian strawflower	α-pinene and γ-curcumene	Antimicrobial and antioxidant	Mollova et al. (2020)
22.	*Calendula officinalis*	Marigold	α-cadinol and Muurolol	Antioxidant and antimicrobial	Ourabia et al. (2019)
23.	*Callistemon viminalis*	Bottle brush	1,8 cineole and α-pinene	Antioxidant and antimicrobial	Salem et al. (2013)
24.	*Melaleuca cajuputi*	Cajeput	Eucalyptol, β-caryophyllene, and α-pinene	Antimicrobial, antioxidant, and antiviral	Isah et al. (2023)
25.	*Elettaria cardamomum*	Cardamom	Eucalyptol, α-pinene, and α-terpineol	Antifungal, antioxidant, and antiulcerogenic	Ashokkumar et al. (2020)

1.5 QUALITY, SAFETY, AND ALLERGENS OF ESSENTIAL OILS AND THEIR FRAGRANCES

Due to lesser adverse effects, EOs and their fragrances have gained popularity in cosmetic formulations and personal care products. Standardization, safety, and allergy evaluation of EOs are important parameters before the commercialization of EOs-based products. Assessment of purity, composition, and sensory qualities of EOs is necessary to determine their quality. Because EOs can be degraded and are volatile, this procedure is complicated. To tackle these problems, several sophisticated analytical methods have been created, each having advantages and disadvantages of its own (Mortimer & Reeder, 2016). Spectroscopic and chromatographic methods are used for the physicochemical characterization of EOs. Other methods like Fourier-transform infrared spectroscopy (FTIR) and gas chromatography-mass spectrometry (GC-MS) are also frequently employed. When used in conjunction with chemometric methods, FTIR can evaluate the purity of EOs and categorize them according to their botanical origin. On the other hand, GC-MS is handy for detecting and quantifying volatile compounds (Agatonovic-Kustrin et al., 2020; Al-Harrasi et al., 2022). Vapor measurement systems like metal-oxide-semiconductor (MOS) gas sensors record and detect the vapor profiles of EOs, offering a rapid and economical way to differentiate the EOs and their compounds (Sudarmaji et al., 2018). Artificial neural networks (ANNs) provide an informal approach by which drying processing parameters may be analyzed to determine the composition and amount of EO. This approach is a potential tool for *in situ* quality evaluation since it accurately predicts quality attributes (Taheri-Garavand et al., 2021). With the chemical composition analysis (CCA), the quality of EOs, and certain bioactive compounds (linalool and linalyl acetate) in lavender oil were determined to be present in specific concentrations. Based on the chemical profiles, cluster analyses, and linear discriminant analyses, a quality assessment was done, and the adulteration of EOs was investigated (Marincaş & Feher, 2018). Although chemical analysis sheds light on the makeup of EOs, sensory quality is not necessarily correlated with it. The volatile profile can be considerably changed by the presence of a plant matrix, indicating that sensory analyses are necessary for an exhaustive quality evaluation. Some novel approaches and techniques, such as the electric conductivity test (ECT), are available to evaluate EOs and their fragrances. The quantity and quality of EO are the main focus of these investigations, which examine the quality of dried plant materials used for EO extraction. This tactic is advantageous when it comes to ensuring the quality of raw resources (Coradi et al., 2018). Using spectrophotometric techniques, one may determine the total phenolic content (TPC) of EOs, which offers information on their antioxidant capabilities. This method helps evaluate the health advantages and caliber of EOs.

Although EOs and their compounds are used in cosmetic, food, and pharma industries, they cause significant toxicity and allergen risks. In addition, considerable exposure to EOs and their compounds causes skin burns due to their photosensitizer properties. Sometimes, they are even linked with dermal reactions and skin dermatitis. Moreover, using cosmetic products with high EO concentrations raises dangers and may cause cutaneous sensitization (Silva, 2004). According to a North American Contact Dermatitis Group research, tea tree oil (*Melaleuca alternifolia*) allergies are thought to affect about 1% of the population. This vulnerability is linked to the relatively high 1,8-cineole concentration (Warshaw et al., 2015). Furthermore, using lavender oil-containing toiletries and cosmetics has been associated with allergic reactions, and a sizable portion of the population (more than 10%) has lavender oil-related allergic contact dermatitis (Crawford et al., 2004). According to Bleasel et al. (2002), some aromatherapists experienced allergic reactions like contact dermatitis to Bulgarian rose essential oils. In another case study, peppermint oil from a lip balm product caused allergic contact cheilitis and lip irritation in some patients. Their lip balm formulation contains lanolin, propolis, coconut oil, almond oil, vitamin E, and peppermint oil. The patch-test findings were positive for peppermint oil, indicating that peppermint oil was the most likely cause of allergic responses (Tran et al., 2010). Some reports suggest that eugenol is hepatotoxic and has significant negative health impacts. In a case study, the eugenol in a cleaning product caused a 34-year-old lady to be diagnosed with dermatitis and occupational asthma when she was using a eugenol mop spray (López Sáez et al., 2015). Menthol and peppermint oil have been detected in skin tests and were subsequently linked to severe anaphylactic events (Bayat & Borici-Mazi, 2014). Even in low quantities, certain EOs may be pretty harmful. Thus, thorough safety evaluations are required for various demographics, including children and expectant mothers. REACH regulation (Regulation (EC) No 1907/2006 of 18 December 2006) refers to the registration, evaluation, authorization, and restriction of chemicals in the European community, leading to the sustainable development of EO products. According to the law, as mandated by the European Community's REACH regulation, EOs must undergo extensive toxicological tests before being produced or imported in considerable quantities. The member nations are accountable for verifying manufacturing principles through ISO 22716 guidelines on good manufacturing practices (GMP), monitoring market information provided by sector operators, testing, and ingredients aimed at allergic sensitives (Barbieri & Borsotto, 2018).

1.6 CONCLUSION

EOs and their aroma compounds are significant in perfume and cosmetics. They provide several advantages for the body and skin and act as natural or phytochemical preservatives. Furthermore, the pleasant smell of these substances adds to the value of cosmetic items. Therefore, the cosmetics and fragrance industries could not exist without these priceless materials. However, it is important to remember that EOs and their constituents may result in allergic responses and symptoms. Cosmetic products that use nanotechnology have better trapping efficiency and dermal penetration of active substances. Furthermore, nanocarrier systems made from aromatic plants satisfy customer demand for environment-friendly goods, encouraging sustainability in the cosmetics industry. Safety assessment of the quantity of EOs, their compounds, and their formulations is required, and the advisory rules must be followed before commercialization.

REFERENCES

Aburjai, T., & Natsheh, F. M. (2003). Plants used in cosmetics. Phytotherapy Research: An International Journal Devoted to Pharmacological and Toxicological Evaluation of Natural Product Derivatives, 17(9), 987–1000.

Agatonovic-Kustrin, S., Ristivojevic, P., Gegechkori, V., Litvinova, T. M., & W. Morton, D. (2020). Essential oil quality and purity evaluation via FT-IR spectroscopy and pattern recognition techniques. Applied Sciences, 10(20), 7294.

Akram, M., Riaz, M., Munir, N., Akhter, N., Zafar, S., Jabeen, F., & Said Khan, F. (2020). Chemical constituents, experimental and clinical pharmacology of *Rosa damascena*: A literature review. Journal of Pharmacy and Pharmacology, 72(2), 161–174.

Al-Harrasi, A., Bhatia, S., Kaushik, D., Ahmed, M. M., Anwer, K., & Sharma, P. B. (2022). Quality Control of Essential Oils. In Al-Harrasi, A., Bhatia, S., Behl, T., Kaushik, D., Anwer, M., Ahmed, M. M., Sharma, P. B., Kabir, M. T., Mittal, V., & Sharma, A. (Eds.), Role of Essential Oils in the Management of COVID-19 (pp. 169–177). CRC Press.

Anand, A., Khurana, R., Wahal, N., Mahajan, S., Mehta, M., Satija, S., & Khurana, N. (2019). Vanillin: A comprehensive review of pharmacological activities. Plant Archives, 19(2), 1000–1004.

Ashokkumar, K., Murugan, M., Dhanya, M. K., & Warkentin, T. D. (2020). Botany, traditional uses, phytochemistry and biological activities of cardamom [*Elettaria cardamomum* (L.) Maton] – A critical review. Journal of Ethnopharmacology, 246, 112244.

Avola, R., Granata, G., Geraci, C., Napoli, E., Graziano, A. C. E., & Cardile, V. (2020). Oregano (*Origanum vulgare* L.) essential oil provides anti-inflammatory activity and facilitates wound healing in a human keratinocytes cell model. Food and Chemical Toxicology, 144, 111586.

Babaei, M., Hesari, A. K., & Soltani, S. (2021). Evaluation of the camphor effects on histological parameters of skin in adult mice and the protective role of vitamin E. Journal of Basic Research in Medical Sciences, 8(2), 53–62.

Barbieri, C., & Borsotto, P. (2018). Essential oils: Market and legislation. In H. A. El-Shemy (Ed.), Potential of Essential Oils (pp. 107–127). IntechOpen.

Batista, P. A., de Paula Werner, M. F., Oliveira, E. C., Burgos, L., Pereira, P., da Silva Brum, L. F., ... & Santos, A. R. (2010). The antinociceptive effect of (-)-linalool in models of chronic inflammatory and neuropathic hypersensitivity in mice. The Journal of Pain, 11(11), 1222–1229.

Bauer, K., Garbe, D., & Surburg, H. (2008). Common Fragrance and Flavor Materials: Preparation, Properties and Uses. John Wiley & Sons.

Bayat, R., & Borici-Mazi, R. (2014). A case of anaphylaxis to peppermint. Allergy, Asthma & Clinical Immunology, 10, 1–3.

Bekka-Hadji, F., Bombarda, I., Djoudi, F., Bakour, S., & Touati, A. (2022). Chemical composition and synergistic potential of *Mentha pulegium* L. and *Artemisia herba alba* Asso. essential oils and antibiotic against multi-drug resistant bacteria. Molecules, 27(3), 1095.

Belwal, T., Ezzat, S. M., Rastrelli, L., Bhatt, I. D., Daglia, M., Baldi, A., & Atanasov, A. G. (2018). A critical analysis of extraction techniques used for botanicals: Trends, priorities, industrial uses and optimization strategies. TrAC Trends in Analytical Chemistry, 100, 82–102.

Bhowal, M., & Gopal, M. (2015). Eucalyptol: Safety and pharmacological profile. Journal of Pharmaceutical Sciences, 5, 125–131.

Bijauliya, R. K., Alok, S., Kumar, M., Chanchal, D. K., & Yadav, S. (2017). A comprehensive review on herbal cosmetics. International Journal of Pharmaceutical Sciences and Research, 8(12), 4930–4949.

Bilal, M., & Iqbal, H. M. (2019). An insight into toxicity and human-health-related adverse consequences of cosmeceuticals—a review. Science of the Total Environment, 670, 555–568.

Bleasel, N., Tate, B., & Rademaker, M. (2002). Allergic contact dermatitis following exposure to essential oils. Australasian Journal of Dermatology, 43(3), 211–213.

Borgonetti, V., López, V., & Galeotti, N. (2022). Ylang-ylang (*Cananga odorata* (Lam.) Hook. f. & Thomson) essential oil reduced neuropathic-pain and associated anxiety symptoms in mice. Journal of Ethnopharmacology, 294, 115362.

Bourgou, S., Rahali, F. Z., Ourghemmi, I., & Saïdani Tounsi, M. (2012). Changes of peel essential oil composition of four Tunisian citrus during fruit maturation. The Scientific World Journal, 2012(1), 528593.

Bouzabata, A., Cabral, C., Gonçalves, M. J., Cruz, M. T., Bighelli, A., Cavaleiro, C., & Salgueiro, L. (2015). *Myrtus communis* L. as source of a bioactive and safe essential oil. Food and Chemical Toxicology, 75, 166–172.

Caissard, J. C., Joly, C., Bergougnoux, V., Hugueney, P., Mauriat, M., & Baudino, S. (2004). Secretion mechanisms of volatile organic compounds in specialized cells of aromatic plants. Recent Research Developments in Cell Biology, 2, 1–15.

Chahal, K. K., Kaushal, S., & Sandhu, A. K. (2015). Chemical composition and biological properties of *Chrysopogon zizanioides* (L.) Roberty syn. *Vetiveria zizanioides* (L.) Nash-a review. Indian Journal of Natural Products and Resources (IJNPR) [Formerly Natural Product Radiance (NPR)], 6(4), 251–260.

Chakraborty, A., Sankaran, V., Ramar, M., & Chellappan, D. R. (2015). Chemical analysis of leaf essential oil of *Cinnamomum verum* from Palni hills, Tamil Nadu. Tic, 3(e10), x10.

Chappell, J. (1995). Biochemistry and molecular biology of the isoprenoid biosynthetic pathway in plants. Annual Review of Plant Biology, 46(1), 521–547.

Chaudhary, A., Kaur, P., Singh, B., & Pathania, V. (2009). Chemical composition of hydrodistilled and solvent volatiles extracted from woodchips of Himalayan Cedrus: *Cedrus deodara* (Roxb.) Loud. Natural Product Communications, 4(9), 1934578X0900400920.

Chaudhri, S. K., & Jain, N. K. (2009). History of cosmetics. Asian Journal of Pharmaceutics (AJP), 3(3), 164–167.

Chauiyakh, O., El Fahime, E., Aarabi, S., Ninich, O., Bentata, F., Kettani, K., & Ettahir, A. (2023). A systematic review on chemical composition and biological activities of cedar oils and extracts. Research Journal of Pharmacy and Technology, 16(8), 3875–3883.

Cheng, H. M., Chen, F. Y., Li, C. C., Lo, H. Y., Liao, Y. F., Ho, T. Y., & Hsiang, C. Y. (2017). Oral administration of vanillin improves imiquimod-induced psoriatic skin inflammation in mice. Journal of Agricultural and Food Chemistry, 65(47), 10233–10242.

Coradi, P. C., Müller, A., Schmidt, D. A., Lima, R. E., Baio, F. H. R., Borsato, A. V., & Blanco, M. (2018). Electric conductivity test for quality assessment of aromatic and medicinal plants after drying. Drying Technology, 36(5), 545–556.

Cortés-Rojas, D. F., de Souza, C. R. F., & Oliveira, W. P. (2014). Clove (*Syzygium aromaticum*): A precious spice. Asian Pacific Journal of Tropical Biomedicine, 4(2), 90–96.

Crawford, G. H., Katz, K. A., Ellis, E., & James, W. D. (2004). Use of aromatherapy products and increased risk of hand dermatitis in massage therapists. Archives of Dermatology, 140(8), 991–996.

da Silva Ramos, R., Rodrigues, A. B. L., Farias, A. L. F., Simões, R. C., Pinheiro, M. T., Ferreira, R. M. D. A., & de Almeida, S. S. M. D. S. (2017). Chemical composition and in vitro antioxidant, cytotoxic, antimicrobial, and larvicidal activities of the essential oil of *Mentha piperita* L. (Lamiaceae). The Scientific World Journal, 2017(1), 4927214.

de, M., Moreira, D., de, L., Valentim, J. T., Ferreira, N., Rodrigues, M., & Cléa, F. (2024). In vivo stress model points (-)-α-bisabolol as an alternative to relieve effects on neuropsychiatric disorders. Archives of Psychiatry and Psychotherapy, 26(2), 45–57. https://doi.org/10.12740/app/175726

de Oliveira Lima, M. I., De Medeiros, A. A., Silva, K. S., Cardoso, G. N., de Oliveira Lima, E., & de Oliveira Pereira, F. (2017). Investigation of the antifungal potential of linalool against clinical isolates of fluconazole resistant *Trichophyton rubrum*. Journal de Mycologie Medicale, 27(2), 195–202.

Desai, M., Parikh, J., & Parikh, P. A. (2010). Extraction of natural products using microwaves as a heat source. Separation & Purification Reviews, 39(1–2), 1–32.

Dilworth, L. L., Riley, C. K., & Stennett, D. K. (2024). Plant Constituents: Carbohydrates, Oils, Resins, Balsams, and Plant Hormones. In McCreath, S. B. & Clement, Y. N. (Eds.), Pharmacognosy (pp. 49–74). Academic Press.

Dobreva, A., & Nedeltcheva-Antonova, D. (2023). Comparative chemical profiling and citronellol enantiomers distribution of industrial-type rose oils produced in China. Molecules, 28(3), 1281.

Dornic, N., Roudot, A. C., Batardière, A., Nedelec, A. S., Bourgeois, P., Hornez, N., & Ficheux, A. S. (2018). Aggregate exposure to common fragrance compounds: Comparison of the contribution of essential oils and cosmetics using probabilistic methods and the example of limonene. Food and Chemical Toxicology, 116, 77–85.

Draelos, Z. D. (2000). Cosmetics and skin care products: A historical perspective. Dermatologic Clinics, 18(4), 557–559.

El Asbahani, A., Miladi, K., Badri, W., Sala, M., Addi, E. A., Casabianca, H., & Elaissari, A. (2015). Essential oils: From extraction to encapsulation. International Journal of Pharmaceutics, 483(1–2), 220–243.

Fejer, J. and Salamon, I. (2016). Breeding of German chamomile, *Matricaria recutita* L., with a high content of **α**-bisabolol. Acta Horticulturae, 1125, 287–292

FEMA (1994) Monograph 2309: DL-Citronellol (FEMA No. 2309), Flavor and Extract Manufacturers' Association, p. 4.

Ganjewala, D. (2009). Cymbopogon essential oils: Chemical compositions and bioactivities. International Journal of Essential Oil Therapeutics, 3(2–3), 56–65.

Gaspar, A. L., Gaspar, A. B., Contini, L. R., Silva, M. F., Chagas, E. G., Bahú, J. O., & Yoshida, C. M. (2022). Lemongrass (*Cymbopogon citratus*)-incorporated chitosan bioactive films for potential skincare applications. International Journal of Pharmaceutics, 628, 122301.

Golda, J. O. (2022). Perfume extraction and formulation from lemongrass leaves. Iconic Research and Engineering Journals, 5(11), 22–33.

Gomes-Carneiro, M. R., Dias, D. M., De-Oliveira, A. C. A. X., & Paumgartten, F. J. (2005). Evaluation of mutagenic and antimutagenic activities of α-bisabolol in the Salmonella/microsome assay. Mutation Research/Genetic Toxicology and Environmental Mutagenesis, 585(1–2), 105–112.

Gowhar, S. A. (2018). Harmful effects of beauty care products on human health. International Journal of Medical Science and Public Health, 7(1), 1–9.

Guo, S., Geng, Z., Zhang, W., Liang, J., Wang, C., Deng, Z., & Du, S. (2016). The chemical composition of essential oils from *Cinnamomum camphora* and their insecticidal activity against the stored product pests. International Journal of Molecular Sciences, 17(11), 1836.

Hammer, K. A., Carson, C. F., & Riley, T. V. (1999). Antimicrobial activity of essential oils and other plant extracts. Journal of Applied Microbiology, 86(6), 985–990.

Han, X., & Parker, T. L. (2017). Anti-inflammatory, tissue remodeling, immunomodulatory, and anticancer activities of oregano (*Origanum vulgare*) essential oil in a human skin disease model. Biochimie Open, 4, 73–77.

Happy, A. A., Jahan, F., & Momen, M. A. (2021). Essential oils: Magical ingredients for skin care. Journal of Plant Sciences, 9(2), 54.

Haynes, W. M. (2014). Lanthanum. Handbook of Chemistry and Physics (Vol. 95, pp. 4–69). CRC Press.

Herman, R. A., Ayepa, E., Shittu, S., Fometu, S. S., & Wang, J. (2019). Essential oils and their applications-a mini review. Advances in Nutrition and Food Science, 4(4), 1–13.

Holopainen, J. K. (2004). Multiple functions of inducible plant volatiles. Trends in Plant Science, 9(11), 529–533.

Horton, M. (2024). Extraction of Sandalwood Essential Oil: Isolating α/3-Santalol. A presentation accessed on 23 July 2024 from https://dc.etsu.edu/boland/2024/presentations/66

Ibrahim, M., Ankwai, G. E., Taave, P., Gungshik, J. R., & Wapwera, J. A. (2021). Optimization of essential oil extraction from *Cymbopogon* spp (lemon grass) and its application in perfume formulation. Journal of Chemical Society of Nigeria, 46(6), 972–977.

Isah, M., Rosdi, R. A., Abdullah, H., Sul'ain, M. D., & Ishak, W. R. W. (2023). Phytoconstituents and biological activities of *Melaleuca cajuputi* Powell: A scoping review. Journal of Applied Pharmaceutical Science, 13(1), 010–023.

Jamir, Y., Bhushan, M., Sanjukta, R., & Robindro Singh, L. (2024). Plant-based essential oil encapsulated in nanoemulsions and their enhanced therapeutic applications: An overview. Biotechnology and Bioengineering, 121(2), 415–433.

Jasuja, N. D., Sharma, S., Choudhary, J., & Joshi, S. C. (2015). Essential oil and important activities of *Thuja orientalis* and *Thuja occidentalis*. Journal of Essential Oil Bearing Plants, 18(4), 931–949.

Kačániová, M., Vukic, M., Vukovic, N. L., Čmiková, N., Verešová, A., Schwarzová, M., & Babošová, M. et al. (2023). An in-depth study on the chemical composition and biological effects of *Pelargonium graveolens* essential oil. Foods, 13(1), 33.

Kamatou, G. P., Vermaak, I., Viljoen, A. M., & Lawrence, B. M. (2013). Menthol: A simple monoterpene with remarkable biological properties. Phytochemistry, 96, 15–25.

Karaca, N., Şener, G., Demirci, B., & Demirci, F. (2021). Synergistic antibacterial combination of *Lavandula latifolia* Medik. essential oil with camphor. Zeitschrift für Naturforschung C, 76(3–4), 169–173.

Kaur, H., Bhardwaj, U., & Kaur, R. (2021a). *Cymbopogon nardus* essential oil: A comprehensive review on its chemistry and bioactivity. Journal of Essential Oil Research, 33(3), 205–220.

Kaur, V., Kaur, R., & Bhardwaj, U. (2021b). A review on dill essential oil and its chief compounds as natural biocide. Flavour and Fragrance Journal, 36(3), 412–431.

Khatri, P. K., Paolini, M., Larcher, R., Ziller, L., Magdas, D. A., Marincas, O., & Bontempo, L. (2023). Botanical characterization and authentication of lavender essential oil using its volatile organic compounds and compound-specific carbon and hydrogen isotope ratio analysis. Food Control, 154, 110002.

Kholibrina, C. R., & Aswandi, A. (2021, June). Application of styrax essential oils in healthy and beauty soap products. In IOP Conference Series: Earth and Environmental Science (Vol. 782, No. 3, p. 032025). IOP Publishing.

Kim, K. Y., Jang, H. H., Lee, S. N., Kim, Y. S., & An, S. (2018). Effects of the myrtle essential oil on the acne skin—clinical trials for Korean women. Biomedical Dermatology, 2, 1–10.

Kim, S., Lee, J., Jung, E., Huh, S., Park, J. O., Lee, J. W., & Park, D. (2008). Mechanisms of depigmentation by α-bisabolol. Journal of Dermatological Science, 52(3), 219–222.

Kivrak, Ş, Göktürk, T., & Kivrak, I (2017). Assessment of volatile oil composition, phenolics and antioxidant activity of bay (*Laurus nobilis*) leaf and usage in cosmetic applications. International Journal of Secondary Metabolite, 4(2), 148–161.

Kumar, K. S., Vani, M. G., & Wang, S. Y. (2022). Limonene protects human skin keratinocytes against UVB-induced photodamage and photoaging by activating the Nrf2-dependent antioxidant defense system. Environmental Toxicology, 37(12), 2897–2909.

Kusuma, H. S., & Mahfud, M. (2017). Comparison of kinetic models of oil extraction from sandalwood by microwave-assisted hydrodistillation. International Food Research Journal, 24(4), 1697–1702.

Langenheim, J. H. (1994). Higher plant terpenoids: A phytocentric overview of their ecological roles. Journal of Chemical Ecology, 20, 1223–1280.

Lapczynski, A., Bhatia, S. P., Letizia, C. S., & Api, A. M. (2008). Fragrance material review on l-citronellol. Food and Chemical Toxicology, 46(11), S110–S113.

Letizia, C. S., Cocchiara, J., Lalko, J., & Api, A. M. (2003). Fragrance material review on linalool. Food and Chemical Toxicology, 41(7), 943–964.

Lide, D. R. (Ed.). (2004). CRC Handbook of Chemistry and Physics (Vol. 85). CRC press.

Linstorm, P. (1998). NIST chemistry webbook, NIST standard reference database number 69. Journal of Physical and Chemical Reference Data, Monograph, 9, 1–1951.

Liu, Y., Yang, L., Zu, Y., Zhao, C., Zhang, L., Zhang, Y., & Wang, W. (2012). Development of an ionic liquid-based microwave-assisted method for simultaneous extraction and distillation for determination of proanthocyanidins and essential oil in *Cortex cinnamomi*. Food Chemistry, 135(4), 2514–2521.

López Sáez, M. P., Carrillo, P., Huertas, A. J., Fernández Nieto, M., & Lopez, J. D. (2015). Occupational asthma and dermatitis induced by eugenol in a cleaner. Journal of Investigational Allergology and Clinical Immunology, 25(1), 64–65.

Lucchesi, M. E., Chemat, F., & Smadja, J. (2004). Solvent-free microwave extraction of essential oil from aromatic herbs: Comparison with conventional hydro-distillation. Journal of Chromatography A, 1043(2), 323–327.

Mahboubi, M. (2019). *Zingiber officinale* Rosc. essential oil, a review on its composition and bioactivity. Clinical Phytoscience, 5(1), 1–12.

Makuch, E., Nowak, A., Günther, A., Pełech, R., Kucharski, Ł, Duchnik, W., & Klimowicz, A. (2020). An advanced mode of antioxidants delivery system effective in the treatment of ROS and skin permeation of the cosmetic formulations-the possible application of clove oil, eugenol and new eugenol derivative. Preprint. https://www.researchgate.net/publication/346203649

Malik, S. (Ed.). (2019). Essential Oil Research: Trends in Biosynthesis, Analytics, Industrial Applications and Biotechnological Production. Springer.

Marincaş, O., & Feher, I. (2018). A new cost-effective approach for lavender essential oils quality assessment. Industrial Crops and Products, 125, 241–247.

Massaldi, H. A., & King, C. J. (1973). Simple technique to determine solubilities of sparingly soluble organics. Solubility and activity coefficients of d-limonene, butylbenzene, and n-hexyl acetate in water and sucrose solutions. Journal of Chemical and Engineering Data, 18(4), 393–397.

Medeleanu, M. L., Fărcaş, A. C., Coman, C., Leopold, L., Diaconeasa, Z., & Socaci, S. A. (2023). Citrus essential oils–Based nano-emulsions: Functional properties and potential applications. Food Chemistry: X, 20, 100960.

Micić, D., Đurović, S., Riabov, P., Tomić, A., Šovljanski, O., Filip, S., & Blagojević, S. (2021). Rosemary essential oils as a promising source of bioactive compounds: Chemical composition, thermal properties, biological activity, and gastronomical perspectives. Foods, 10(11), 2734.

Miller, J. A., Thompson, P. A., Hakim, I. A., Lopez, A. M., Thomson, C. A., Chew, W., & Chow, H. H. S. (2012). Safety and feasibility of topical application of limonene as a massage oil to the breast. Journal of Cancer Therapy, 3(5A), 749–754.

Mohammad Azmin, S. N. H., Abdul Manan, Z., Wan Alwi, S. R., Chua, L. S., Mustaffa, A. A., & Yunus, N. A. (2016). Herbal processing and extraction technologies. Separation & Purification Reviews, 45(4), 305–320.

Mohammed, H. H., Laftah, W. A., Ibrahim, A. N., & Yunus, M. A. C. (2022). Extraction of essential oil from *Zingiber officinale* and statistical optimization of process parameters. RSC Advances, 12(8), 4843–4851.

Mollova, S., Fidan, H., Antonova, D., Bozhilov, D., Stanev, S., Kostova, I., & Stoyanova, A. (2020). Chemical composition and antimicrobial and antioxidant activity of *Helichrysum italicum* (Roth) G. Don subspecies essential oils. Turkish Journal of Agriculture and Forestry, 44(4), 371–378.

Mortimer, S., & Reeder, M. (2016). Botanicals in dermatology: Essential oils, botanical allergens, and current regulatory practices. Dermatitis, 27(6), 317–324.

Nakahara, K., Alzoreky, N. S., Yoshihashi, T., Nguyen, H. T., & Trakoontivakorn, G. (2013). Chemical composition and antifungal activity of essential oil from *Cymbopogon nardus* (citronella grass). Japan Agricultural Research Quarterly: JARQ, 37(4), 249–252.

Nardelli, A., Drieghe, J., Claes, L., Boey, L., & Goossens, A. (2011). Fragrance allergens in 'specific' cosmetic products. Contact Dermatitis, 64(4), 212–219.

Navarra, M., Mannucci, C., Delbò, M., & Calapai, G. (2015). *Citrus bergamia* essential oil: From basic research to clinical application. Frontiers in Pharmacology, 6, 36.

Nikfar, S., & Behboudi, A. F. (2014). Limonene. In P. Wexler (Ed.), encyclopedia of toxicology (3rd ed Vol. 3, pp. 78–81). Academic Press.

Nikolić, M., Glamočlija, J., Ferreira, I. C., Calhelha, R. C., Fernandes, Â. Marković, T., & Soković, M. (2014). Chemical composition, antimicrobial, antioxidant and antitumor activity of *Thymus serpyllum* L., *Thymus algeriensis* Boiss. and Reut and *Thymus vulgaris* L. essential oils. Industrial Crops and Products, 52, 183–190.

Ourabia, I., Djebbar, R., Tata, S., Sabaou, N., & Fouial-Djebbar, D. (2019). Determination of essential oil composition, phenolic content, and antioxidant, antibacterial and antifungal activities of marigold (*Calandula officinalis* L.) cultivated in Algeria. Carpathian Journal of Food Science & Technology, 11(2), 93–110.

Padole, V. N., Sarade, S., Rathod, S., More, S., & Mendhi, S. (2022). Multivalent role of essential oil in cosmetics: A review. International Journal of Pharmaceutical Sciences Review and Research, 73, 97–105.

Patel, T., Ishiuji, Y., & Yosipovitch, G. (2007). Menthol: A refreshing look at this ancient compound. Journal of the American Academy of Dermatology, 57(5), 873–878.

Patgar, B. G., Satish, S., & Shabaraya, A. R. (2021). Essential oil of *Cupressus sempervirens*: A brief review. World Journal of Pharmacy and Pharmaceutical Sciences, 10(4), 864–872.

Patkar, K. (2008). Herbal cosmetics in ancient India. Indian Journal of Plastic Surgery, 41(S 01), 134–137.

Pavithra, P. S., Mehta, A., & Verma, R. S. (2019). Essential oils: From prevention to treatment of skin cancer. Drug Discovery Today, 24(2), 644–655.

Perbellini, L., Gottardo, R., Caprini, A., Bortolotti, F., Mariotto, S., & Tagliaro, F. (2004). Determination of alpha-bisabolol in human blood by micro-HPLC–ion trap MS and head space-GC–MS methods. Journal of Chromatography B, 812(1–2), 373–377.

Peris, I., & Blázquez, M. A. (2015). Comparative GC-MS analysis of bay leaf (*Laurus nobilis* L.) essential oils in commercial samples. International Journal of Food Properties, 18(4), 757–762.

Poonkodi, K. (2016). Chemical composition of essential oil of *Ocimum basilicum* l.(basil) and its biological activities-an overview. Journal of Critical Review, 3(3), 56–62.

Price, S., Price, L., & Price, P. (2021). Aromatherapy for health professionals: Revised reprint (5th ed.). Elsevier Health Sciences.

Rather, M. A., Dar, B. A., Sofi, S. N., Bhat, B. A., & Qurishi, M. A. (2016). Foeniculum vulgare: A comprehensive review of its traditional use, phytochemistry, pharmacology, and safety. Arabian Journal of Chemistry, 9, S1574–S1583.

Rauber, C. D. S., Guterres, S. S., & Schapoval, E. E. (2005). LC determination of citral in *Cymbopogon citratus* volatile oil. Journal of Pharmaceutical and Biomedical Analysis, 37(3), 597–601.

Rehman, A., Iqbal, M., Khan, B. A., Khan, M. K., Huwaimel, B., Alshehri, S., & Hosny, K. M. (2022). Fabrication, in vitro, and in vivo assessment of eucalyptol-loaded nanoemulgel as a novel paradigm for wound healing. Pharmaceutics, 14(9), 1971.

Rehman, R., Hanif, M. A., Mushtaq, Z., & Al-Sadi, A. M. (2016). Biosynthesis of essential oils in aromatic plants: A review. Food Reviews International, 32(2), 117–160.

Rozza, A. L., Beserra, F. P., Vieira, A. J., Oliveira de Souza, E., Hussni, C. A., Martinez, E. R. M., & Pellizzon, C. H. (2021). The use of menthol in skin wound healing—Anti-inflammatory potential, antioxidant defense system stimulation and increased epithelialization. Pharmaceutics, 13(11), 1902.

Rubiolo, P., Sgorbini, B., Liberto, E., Cordero, C., & Bicchi, C. (2010). Essential oils and volatiles: Sample preparation and analysis. A review. Flavour and Fragrance Journal, 25(5), 282–290.

Salem, M. Z., Ali, H. M., El-Shanhorey, N. A., & Abdel-Megeed, A. (2013). Evaluation of extracts and essential oil from *Callistemon viminalis* leaves: Antibacterial and antioxidant activities, total phenolic and flavonoid contents. Asian Pacific Journal of Tropical Medicine, 6(10), 785–791.

Samanta, S. (2021). Impacts of chemical beauty: Impacts of chemical beauty. Indian Journal of Physiology and Allied Sciences, 73(1), 1–6.

Sangwan, N. S., Farooqi, A. H. A., Shabih, F., & Sangwan, R. S. (2001). Regulation of essential oil production in plants. Plant Growth Regulation, 34, 3–21.

Sareriya, K. J., & Vanzara, P. B., & Upadhyay, D. (2023) Methodology For extraction of essential oils: A review. International Journal for Multidisciplinary Research, 5(4), 1–26.

Satyal, P., & Setzer, W. N. (2020). Chemical compositions of commercial essential oils from *Coriandrum sativum* fruits and aerial parts. Natural Product Communications, 15(7), 1934578X20933067.

Schmidt, E., Jirovetz, L., Wlcek, K., Buchbauer, G., Gochev, V., Girova, T., & Geissler, M. (2007). Antifungal activity of eugenol and various eugenol-containing essential oils against 38 clinical isolates of *Candida albicans*. Journal of Essential Oil Bearing Plants, 10(5), 421–429.

Sethunga, M., Ranaweera, K. K. D. S., Gunathilake, K. D. P. P., & Munaweera, I. (2022). Recent advances in the extraction methods of essential oils and oleoresins from plant materials and its potential applications: A comprehensive review. Journal of Food and Bioprocess Engineering, 5(2), 151–167.

Sharifi-Rad, J., Quispe, C., Kumar, M., Akram, M., Amin, M., Iqbal, M., & Cho, W. C. (2022). Hyssopus essential oil: An update of its phytochemistry, biological activities, and safety profile. Oxidative Medicine and Cellular Longevity, 2022(1), 8442734.

Sharma, A., Kumar, V., Mittal, C., Rana, V., Dabral, K., & Parveen, G. (2023). Role of essential oil used pharmaceutical cosmetic product. Journal for Research in Applied Sciences and Biotechnology, 2(3), 147–157.

Shiwakoti, S., Sintim, H. Y., Poudyal, S., Bufalo, J., Cantrell, C. L., Astatkie, T., & Zheljazkov, V. D. (2015). Diurnal effects on *Mentha canadensis* oil concentration and composition at two different harvests. HortScience, 50(1), 85–89.

Sikdar, D. C., Menon, R., Duseja, K., Kumar, P., & Swami, P. (2016). Extraction of citrus oil from orange (*Citrus sinensis*) peels by steam distillation and its characterizations. International Journal of Technical Research and Applications, 4(3), 341–346.

Silva, A. R. D. (2004). Aromaterapia em dermatologia e estética. Roca, p. 432.

Snoussi, M., Noumi, E., Trabelsi, N., Flamini, G., Papetti, A., & De Feo, V. (2015). *Mentha spicata* essential oil: Chemical composition, antioxidant and antibacterial activities against planktonic and biofilm cultures of Vibrio spp. strains. Molecules, 20(8), 14402–14424.

Southwell, I. (2021). *Backhousia citriodora* F. Muell. (lemon myrtle), an unrivalled source of citral. Foods, 10(7), 1596.

Sudarmaji, A., Margiwiyatno, A., Ediati, R., & Mustofa, A. (2018, May). Vapor measurement system of essential oil based on MOS gas sensors driven with advanced temperature modulation technique. In IOP Conference Series: Earth and Environmental Science (Vol. 147, No. 1, p. 012046). IOP Publishing.

Swamy, M. K., & Sinniah, U. R. (2015). A comprehensive review on the phytochemical constituents and pharmacological activities of *Pogostemon cablin* Benth.: An aromatic medicinal plant of industrial importance. Molecules, 20(5), 8521–8547.

Taheri-Garavand, A., Mumivand, H., Fanourakis, D., Fatahi, S., & Taghipour, S. (2021). An artificial neural network approach for non-invasive estimation of essential oil content and composition through considering drying processing factors: A case study in *Mentha aquatica*. Industrial Crops and Products, 171, 113985.

Tongnuanchan, P., & Benjakul, S. (2014). Essential oils: Extraction, bioactivities, and their uses for food preservation. Journal of Food Science, 79(7), R1231–R1249.

Tran, A., Pratt, M., & DeKoven, J. (2010). Acute allergic contact dermatitis of the lips from peppermint oil in a lip balm. DERM, 21(2), 111–115.

Tran, T. H., Nguyen, D. C., Phu, T. N. N., Bach, L. G., & Nguyen, T. D. (2019). Research on lemongrass oil extraction technology (hydrodistillation, microwave-assisted hydrodistillation). Indonesian Journal of Chemistry, 19(4), 1000–1007.

Tripathi, L., Kumar, P., Swain, K., & Pattnaik, S. (2023). Clinical Applications of Essential Oils. In Inamuddin, T. Altalhi, & J. N. Cruz (Eds.), Essential Oils: Extraction Methods and Applications (pp. 933–951). Wiley-Scrivener.

Uronnachi, E., Atuegwu, C., Umeyor, C., Nwakile, C., Obasi, J., Ikeotuonye, C., & Attama, A. (2022). Formulation and evaluation of hair growth enhancing effects of oleogels made from rosemary and cedar wood oils. Scientific African, 16, e01223.

Vollmer, D. L., West, V. A., & Lephart, E. D. (2018). Enhancing skin health: By oral administration of natural compounds and minerals with implications to the dermal microbiome. International Journal of Molecular Sciences, 19(10), 3059.

Wagner, K. H., & Elmadfa, I. (2003). Biological relevance of terpenoids: Overview focusing on mono-, di-and tetraterpenes. Annals of Nutrition and Metabolism, 47(3-4), 95–106.

Wang, M., Chittiboyina, A. G., Avonto, C., Parcher, J. F., & Khan, I. A. (2014). Comparison of current chemical and stereochemical tests for the identification and differentiation of *Pelargonium graveolens* L'Hér. (Geraniaceae) essential oils: Analytical data for (-)-(1S, 4R, 5S)-guaia-6, 9-diene and (-)-(7R, 10S)-10-epi-γ-eudesmol. Records of Natural Products, 8(4), 360–372.

Wang, Y., Li, B., Zhu, L., Wang, P., Xu, F., & Zhang, Y. (2022). Octenyl succinic acid starch-stabilized vanilla essential oil pickering emulsion: Preparation, characterization, antioxidant activity, and storage stability. Foods, 11(7), 987.

Warshaw, E. M., Maibach, H. I., Taylor, J. S., Sasseville, D., DeKoven, J. G., Zirwas, M. J., & Belsito, D. V. (2015). North American contact dermatitis group patch test results: 2011–2012. Dermatitis, 26(1), 49–59.

Wickstrom, E. (1988). Camphor. Oslo: National Poison Center (PIM 095).

Yuwono, M., Hafid, A. F., Poernomo, A. T., Agil, M., Indrayanto, G., & Ebel, S. (2002). Eugenol. In H. G. Brittain (Ed.), Analytical Profiles of Drug Substances and Excipients, (Vol. 29, pp. 149–177). Academic Press.

Chapter 2

Historical journey of essential oils and fragrances used in cosmetics

Essential oils and fragrances used in cosmetics

Tanya Singh Raghuvanshi and Bhanu Prakash
Centre of Advanced Study in Botany, Institute of Science, Banaras Hindu University, Varanasi, India

Priyanka Singh
Department of Botany, Maharaja Suhel Dev State University, Azamgarh, India

2.1 INTRODUCTION

The use of cosmetics has been a significant part of human life for thousands of years. They have been an integral part of the daily regime of a wide population for the protection of skin and the improvement of beauty. Numerous flavors and fragrances have permeated the daily lives of humans over time, most notably in the form of foods, drinks, confectionaries, medicines, skincare, perfumes, and other cosmetics. All cosmetics were created with natural materials in the past, primarily from plants, minerals, and animals. In the 20th century, the widespread use of synthetic and chemically curated substances in cosmetics have been reported in traditional literature (Sharmeen et al., 2021).

Going back to earlier times, the use of natural ingredients in cosmetic formulations began to gain popularity in the 1990s and early 2000s. This trend was probably started by consumer demands and cosmetic manufacturers attempting to stand out in a competitive marketplace. At first, the majority of the natural components were skincare actives, which caused the skin to react biologically; usually to improve the skin's structural integrity, offering sun protection, minimizing photoaging, chronic conditions, and other benefits (McMullen & Dell'Acqua, 2023). Since then, the focus of industries has been on creating more sustainable products that are considered to imbue the benefit to the users without endangering the environment. Several companies in the cosmetics sector were searching for natural components to substitute synthetic ones in formulations. The goal is to use their biological characteristics to create innovative products that improve human wellness, health, and appearance. Thus, investigating the advantages of incorporating botanical extracts into various cosmetic product formulations has been prompted by current market developments and customer interests. Specifically, items that meet

DOI: 10.1201/9781003558477-2

certain of the parameters required in the design of new products are highly valued (Guzmán & Lucia, 2021; Lucia et al., 2017). Among natural products, essential oils (EOs) and their active pharmacological compounds (EOCs) could play an important role in cosmetic products because of their pleasant aroma and antimicrobial and therapeutic ingredients. They are biologically active and act as analgesics, carminatives, diuretics, antiseptics, antimicrobials, spasmolytics, hyperaemic stimulants, and diuretics. They are proven to have skin benefits like anti-aging, anti-wrinkle, skin brightening, skin whitening, anti-acne, and anti-blemish (Raghuvanshi et al., 2023). However, their pleasant aroma is the primary factor behind their use in cosmetics. The fatty acids, fatty oils, and surfactants that are used in the manufacturing of cosmetics frequently have an unpleasant smell; EOs acts as a natural alternative to mask those intense smells (Sarkic & Stappen, 2018).Cosmetic preparations like lotions, gels, and ointments that contain EO or a mixture of EO-derived actives don't always need chemical preservatives because of the antimicrobial properties of EOs. The invention of distillation methods by the Arabs, the spread of the still, and the translation of alchemical treatises by physicians from the School of Salerno in the 12th and 13th centuries made the production of perfumes feasible. The manufacturing of EO s was industrialized in the first half of the 19th century as a result of the huge demand for these oils as ingredients in perfumes, even if distillation methods were refined over the ages, some of the most prominent ones being water or steam distillation, solvent extraction, expression under pressure, supercritical fluid or subcritical water extractions (Burger et al., 2019). Further, the journey of EO with due course of time is thus fascinating and discussed later in this chapter.

2.2 CHEMICAL COMPOSITION AND AROMA PROFILE

EOs, often referred to as volatile or ethereal oils, are naturally occurring volatile organic chemicals that are made in large amounts from raw plant materials or their parts, including wood, roots, bark, twigs, flowers, seeds, buds, leaves, and fruits. These oils are a complex mixture of over 300 different chemicals with molecular weight typically less than 300 Da. Chemically speaking, such substances fall under a category called secondary metabolites of plants including a number of aromatic compounds such as ethers, phenols, terpenes, alcohols, amines, and carbonyl substances including amides, esters, ketones, and aldehydes. The production and composition of EOs in plant are influenced by a number of factors such as physiological variations (organ development, pollinator activity cycle, type of plant material (leaves, root, stem, etc.), type of secretory structure, seasonal variation, mechanical or chemical injuries), environmental conditions (climate, pollution, diseases and pests edaphic factors),geographic variation, genetic factors and evolution, storage, political/social conditions, amount of plant material/space and extraction and purification methods used (Figueiredo et al., 2008). The main chemical components of EOs consist of phenylpropanoids and terpenes. Compounds with one or more chiral carbon atoms have optical activity. These chemicals exist in many enantiomeric forms due to their chirality. There are various uses for these enantiomers and diastereomers in the food, pharmaceutical, fragrance, and flavor industries (Lahlou, 2004). Furthermore, these enantiomers differ in their fragrance characteristics as well as their applications. For example, S-(+)-carvone has an aromatic flavor, but R-(−)-carvone has a minty flavor. Furthermore, compared to the (−) -limonene enantiomer, the scent of the (+)-limonene enantiomer is noticeably more stimulating (Heuberger et al., 2001). Individual enantiomers' potential for therapeutic benefit is influenced by differences in their scent profiles (Laska & Teubner, 1999;Vora et al., 2024).

2.3 SOURCE OF ESSENTIAL OILS

EOs are extracted from different aromatic plants. These plants can be found throughout the Mediterranean and tropical nations. Due to the fact that the locals employ these plants to treat illnesses, they have gained importance in day to day lives. Every part of the plant, including its leaves, seeds, buds, stems, and flowers, produces EO. The epidermic cells, cavities, secretary cells, and channels are where EO accumulates (Gilani et al., 2005). EOs in plants are what give them a particular aroma. Plant materials that are fresh, dried, or partially dehydrated are usually used to extract EOs. The diffusion through plant tissues that directly contact the surface from which the EO are extracted using various techniques determines the extraction rate. The stability of the EO is a prerequisite for its extraction. The steam distillation method and the hydro distillation process are the two most important techniques used to extract EOs. These methods are the most appropriate and productive ones for extraction procedures (Irshad et al., 2020; Kulisic et al., 2004). A few other techniques are also employed for extraction, although they are not equally appropriate for this procedure. These included the use of liquid carbon dioxide or microwaves, as well as high- or low-pressure distillation using hot or steaming water (Bousbia et al., 2009). While EOs used in the fragrance or perfume industry are extracted from lipophilic solvents and occasionally with supercritical carbon dioxide to make them more appealing, EOs extracted from the steam distillation method are primarily used in pharmacological activities and food items. The age of the plants, the portions used for extraction, the stage of the vegetative cycle, the impact of the climate, etc., all had an impact on the quality of the EO. The chemical analysis of EOs is completely transformed by the spectroscopic and chromatographic techniques. The chemical makeup of EOs can be investigated using gas chromatography, NMR spectroscopy, UV-Vis spectroscopy, and IR spectroscopy (Donelian et al., 2009; Hussain et al., 2008). Gas chromatography (GC and GC-MS) is commonly used in the process of characterizing the EO. The quantity and quality of EO are also influenced by how it is handled and stored (Irshad et al., 2020).

2.4 TIMELINE OF USE OF ESSENTIAL OIL FROM ANCIENT TO MODERN TIME

2.4.1 Emergence of essential oil during ancient civilizations

Early accounts describe King Sahure's journey to the legendary Punt, which is thought to be present-day Ethiopia or Somalia, about 2400 BCE. Among other treasures, King Sahure brought back 8000 measures of myrrh. Around 1500 temple pictographs describe Queen Hatshephut's expedition to Punt, with the aim of returning with frankincense trees to be replanted in Egypt. The Greeks took the use of scents a step further, employing them not only to honor their gods but also for hedonistic purposes and in exercises and treatments that marked the inception of early kinds of aromatherapy. Rose, saffron, frankincense, myrrh, violets, spikenard, cinnamon, and cedarwood were the most commonly utilized perfumes by the Greeks, who exported these aromatics widely throughout the Mediterranean and Middle East.

Egyptians were pioneers in the use of EOs for beauty and cosmetics. They used oils such as frankincense, myrrh, and rose in perfumes and skincare. Cleopatra, in particular, was known for using oils in her beauty routines. The Egyptians utilized fragrant herbs for cosmetic and medical uses as early as 3000 BC. Hippocrates, who is still regarded as the founding father of medicine, makes reference to several therapeutic and medicinal plants in his writings. The 10th century saw the Arabic Doctor Avicenna's important written records detailing the impacts of 800 plants on the human body. He is also known for having invented the distillation procedure of EOs (Pujari et al., 2020).

The Chinese utilized EOs for medicinal and cosmetic purposes and in balms, ointments, and perfumes. Ayurveda, one of the world's oldest holistic healing systems, incorporates EOs like sandalwood and jasmine in beauty treatments to purify the skin and calm the mind. Around 1000 BCE, Greeks, influenced by Egyptian practices, used EOs such as lavender and rosemary for their rejuvenating properties. Hippocrates advocated the use of oils for health and wellness, including skincare. Romans popularized the use of fragrant oils for personal hygiene, beauty, and luxurious bathing rituals, using oils like rose and chamomile (Figure 2.1).

The Vedas, which date back around 3000 years, have detailed and informed references to cosmetics and perfumes, which are primarily utilized in religious events. According to Rao and Bhattacharya (1992), items listed in these works from the earliest recorded ages to the time of Kalidasa's life (470 AD) include

Figure 2.1 Timeline representing the journey of essential oils.

the blossoms of henna, jasmine, campak, lotus, mango, hyacinth, rose, lily, sandalwood, and vetiver roots. The first Indian writing was known as Abir and included camphor, patchouli leaves, sandalwood, and vetiver root. EOs have long been used in the 3000-year-old Ayurvedic system of traditional Indian medicine to promote healing. More than 700 ingredients, including cinnamon, ginger, and sandalwood, are listed as beneficial for healing in Vedic literature. During the Bubonic Plague outbreak, Ayurveda was utilized to successfully replace antibiotics that were not working. Aromatic plants and oils had more than just medical purposes; they were also seen as a divine aspect of the natural world and were fundamental to the spiritual and philosophical tenets of Ayurvedic treatment (Pujari et al., 2020).

2.4.2 Change in the use of essential oil during middle ages with development of extraction methods and trade routes

In an unusual manner, the use of aromatic products which had already begun to blossom from Eastern alchemical practice grew rapidly as a result of the Black Death of 1347–1351 and subsequent pandemics. It was thought that breathing contaminated air could spread plague. In order to mask the smell of decay emanating from dead bodies lying in streets, individuals carried nosegays and tiny floral arrangements.

The era of globe discovery reached its peak in the 16th and 17th centuries. The Italian Columbus's expeditions to the Americas, with the Englishman John Cabot pursuing closely after, had preceded it. Cartier also traveled the North American coast (1534–1536) in pursuit of wealth, while Verrazzano looked for a Northern Strait (1524–1528) (Sell & Pybus, 2006).

Magellan and da Gama explored the southern seas, while Queen Elizabeth's master mariners Drake, Gilbert, Frobisher, and Raleigh trawled the waters in search of riches for her treasury. In addition to increasing the supply of already-existing scent items, all of this oceanic activity brought numerous novel aroma products back to Europe. The alchemists were eager to expand their knowledge at the other end of the journey. In the meantime, Paracelsus (1493–1541) furthered the development of the quinta essential (quintessence) notion of a fifth element, implicated in imbuing life, and worked on distillation to separate the "essential" from the "non-essential" portions of a compound. His unconventional theories had a significant impact on Renaissance medicine and thus the use of EO. Paracelsus believed that although God makes medicine, it is not in its usable form. Medicine is present in nature mixed with "dross," which needs to be removed via distillation in order to liberate the medication. At this time, court perfumers began to appear and the first books and manuscripts detailing perfumery procedures appeared. Driven by the English Civil War (1642–1651), the opposition to monarchy intensified in the North American and French revolutions, resulting in a radically altered political landscape and global order. A more subdued revolution was taking place in the interim as chemists, spicers, and apothecaries started to professionally specialize in a new trade: perfumery. A groundbreaking fragrance comprising orange flower, musk, civet, violet, and amber was launched by London perfumer Charles Lilly in 1708, and William Bayley established a perfumery in Long Acre in 1711 (Naeem et al., 2018; Sell & Pybus, 2006).

In an effort to classify the spectrum of smells, S. Piesse developed theories in his book *Odours, Fragrances and Cosmetics* (1879) that connected particular smells to notes on a musical scale. Atkinsons, on the other hand, wrote one of the first books on perfume technology in 1890, focusing primarily on the cryoscopic removal of fats to produce absolutes. The House of Worth made the costumes for Empress Eugenie, the powerful spouse of Napoleon III, for whom Guerlain developed Eau Imperiale in 1861. It was discovered at the end of the century that the reformulated scent was centered on neroli, rose, geranium, sandalwood, musk, and the artificial chemical coumarin. With the introduction of top-, middle-, and bottom-note terms, fragrances started to be described in a structural format. Science, industrialization, consumer demand, and the people of the time came together toward the end of the 19th century to create a catalyst for the rise in the use of smell and the enjoyment it brought. At last, perfume was becoming widely available. This established the framework for the 20th century, the era of fashion, which sparked an incredibly rapid expansion in the variety of applications for fragrance. Amid the bustle of activity surrounding perfumes, significant advancements had been made in the synthesis of aroma compounds, which significantly impacted and helped formulation chemists and perfumers in their selection and the selection of other materials (Figure 2.2) (Sell & Pybus, 2006).

2.4.3 Modern-day utilization of essential oil in beauty and cosmetics and role of science and technology

There has been a surge in the cosmetic industry's efforts to make sustainable cosmetics using natural ingredients in the late 20th and early 21st centuries (Figure 2.3). Products made from naturally occurring polysaccharides found in cellulose, chitosan, guar gum, starch, lipids, and a variety of other substances marked the beginning of this movement (Apone et al., 2019; Dini & Laneri, 2021; Morocho-Jácome et al., 2021; Shivanand et al., 2010). These are the functional biopolymers in the formulation that act as film formers, emulsifiers, binding agents, and rheology modifiers. Significant progress has been made in the field of skin care active ingredients in recent years. This is

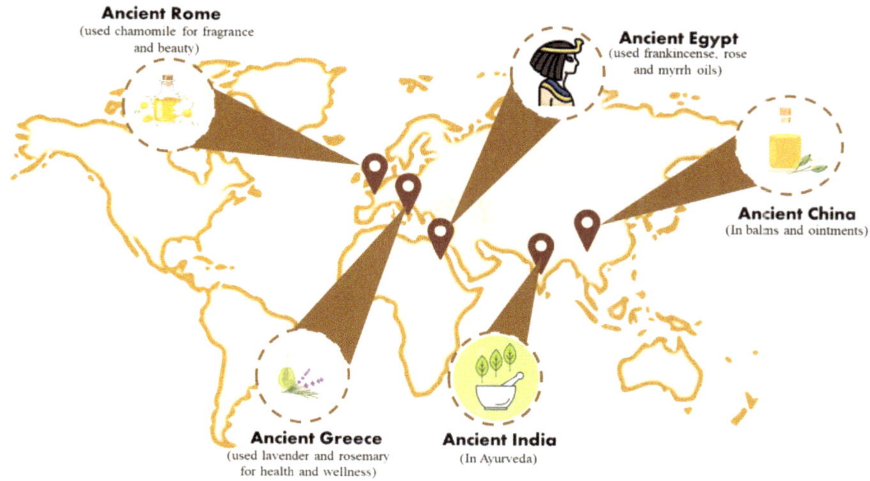

Figure 2.2 Use of essential oil dating back to ancient civilizations.

classified as cosmeceuticals and usually entails peptide and botanical component intervention in the dermis and epidermis. In the end, effective treatments increase skin hydration and whitening or reduce the signs of aging on skin. In the past 20 years, a lot of progress has been made in the field of skin care technologies pertaining to antioxidants, which are present in botanical components (McMullen, 2018). Antioxidants are used in practically all skin care

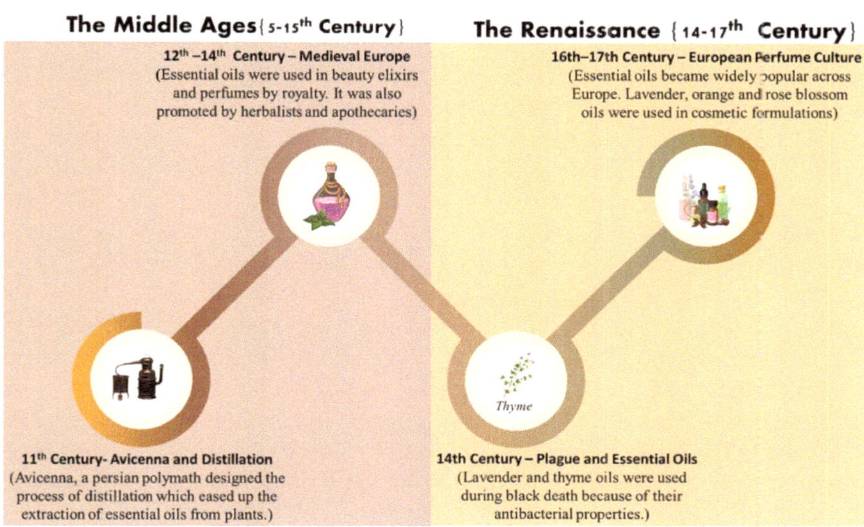

Figure 2.3 Development and changes through middle ages affecting cosmetics.

products today, either to avoid oxidation in the cosmetic formula or to get the antioxidant to the skin so it can operate biochemically. There is a growing demand for consumer product companies in the personal care sector to create environmentally friendly, sustainable products. As a result, a significant portion of the formula's constituents must come from natural sources, according to product development programs. Nevertheless, achieving the same degree of performance as conventional components presents numerous obstacles. The development of eco-friendly preservatives, surfactants, hair styling agents, sunscreens, and many other substances is still a major focus in the personal care product industry (McMullen &Dell'Acqua, 2023).

2.4.4 Demand for essential oils in cosmetic industries in the 21st century

Synthetic fragrances, created in laboratories to replicate the aromatic and chemical components of natural, plant-based oils that are costly to produce, are used in a wide range of businesses. However, synthetic perfumes could not have the health benefits of actual EOs derived from plants and might even be dangerous to use on humans. For example, phthalates, which are recognized carcinogens like benzene derivatives and endocrine disruptors, are compounds included in artificial perfumes. However, the use of natural fragrances, like EOs, over synthetic ones is growing due to their many health benefits, including those associated with aromatherapy. This is expected to fuel the growth of the market for natural fragrances globally (Çeçen and Tezel, 2017; Olujimi et al., 2010). The size of the global market for EOs, for example, was estimated to be more than USD 7.51 billion in 2018 and is predicted to increase at a compound annual growth rate (CAGR) of more than 9% between 2019 and 2026. EOs, like the orange oil used in cosmeceuticals, have been very profitable. This is because they provide a variety of health benefits, including firmness and elasticity to the skin, as well as use in the treatment for acne, scars, and stretch marks. Due to their antibacterial qualities, lemon and orange EOs are excellent choices for skin and hair care products (Table 2.1). Additionally, these oils' antibacterial qualities have created new opportunities for corporate expansion. Large volumes of certain EOs are produced; for example, in 2008, almost 51,000 tons of orange oils, 32,000 tons of maize mint oils, and 9200 tons of lemon oils were produced. The most widely used natural EOs are derived from a range of citrus fruits, and they make up the majority of commercial natural flavors and perfumes. However, because of their scarcity, several others like iris (6200–100,000 €/kg depending on iron content), rose oil (6000–10,000 €/kg), and agarwood oil (6000–11,000 €/kg) are produced on a much smaller scale and traded at extremely high prices. These prices fluctuate and could be influenced by factors like extraction yield, climate dependence, problems with harvesting, or raw material scarcity (Adlard, 2010; Baser & Buchbauer, 2009; Sawamura, 2011).

Table 2.1 Selected essential oils and their topical benefits

S.no.	Essential oil	Plant part	Major active constituents	Properties/Use	Reference
1.	*Citrus bergamia*	Fruit	Limonene	Alleviating acne vulgaris	Navarra et al. (2015) Sun et al. (2020)
2.	*Rosmarinus officinalis*	Leaves	Eucalyptol (1,8-cineole)	Antioxidant, anti-hair loss, promotes new hair growth	Ezekwe et al. (2020) Hussain et al. (2010)
3.	*Mentha piperita*	Leaves	Menthol	Antidandruff, act against inflammation, eczema, hives, psoriasis	Zhao et al. (2022)
4.	*Lavandula angustifolia*	Flower buds	Linalyl acetate, linalool	Hair growth, antimicrobial activity, anti-inflammatory, anti-acne	Kozuharova et al. (2023) Prusinowska & Śmigielski (2014)
5.	*Rosa canina*	Fruits	β-caryophyllene, geraniol	Anti-aging, anti-wrinkle, moisturizing	Ghazghazi et al. (2012) Phetcharat et al. (2015)
6.	*Pogostemoncablin*	Leaves	Patchoulol	Anti-wrinkle, moisturizing	Kusuma & Mahfud (2017) Rahmi et al. (2021)
7.	*Cananga odorata*	Flowers	β-caryophyllene	Skin hydration, anti-wrinkle	Alam et al. (2024) Giang & Son (2016)
8.	*Syzygium aromaticum*	Leaves and buds	Eugenol	Antioxidant, anti-aging, anti-wrinkle	Huang et al. (2013)
9.	*Cymbopogon nardus*	Leaves	Citronellal	Anti-acne, anti-aging	Huang et al. (2013)
10.	*Santalum spicatum*	Heartwood	Lanceol and nuciferol	Anti-acne, used to treat psoriasis, eczema, common warts, and molluscum contagiosum	Moy & Levenson (2017)
				Moisturizing, soothing	

(*Continued*)

Table 2.1 (Continued) Selected essential oils and their topical benefits

S.no.	*Essential oil*	*Plant part*	*Major active constituents*	*Properties/Use*	*Reference*
11.	*Myristica fragrans*	dried aril	Elemicin	Anti-aging, anti-wrinkle	Al-Qahtani et al. (2022) Rahmi et al. (2021)
12.	*Citrus sinensis*	Fruit peel	Limonene and Myrcene	Oil control, pore minimizing, anti-wrinkle	Ao et al. (2008) Cholke et al. (2017)
13.	*Oenothera biennis*	Seeds	β-amyrin	Anti-wrinkles, moisturizing, anti-acne	Muggli (2005) Parham et al. (2020)
14.	*Salvia officinalis*	Leaves	α-terpineol	Antioxidant, anti-aging	Ghorbani & Esmaeilizadeh (2017) Parham et al. (2020)

Additionally, well-known cosmetic manufacturers incorporate a variety of EOs into their products; the most costly perfumes, on the other hand, are made with pure EOs, which give them distinctive scents and set them apart as premium fragrances. Despite being beneficial for health and appearance,, there may be some instances in which EOs should not be used due to allergies or contraindications (Butnariu & Sarac, 2018; Sarkic & Stappen, 2018).

Extensive commercial harvesting of aromatic plants puts certain species at risk of extinction due to lack of habitat. Therefore, it is essential to implement systematic cultivation of these plants to preserve biodiversity and safeguard endangered species. Agro-based solvents, process intensification, sustainable natural product extraction, and manufacturing technologies are the keys to creating environmentally conscious aromatic compounds that span the entire olfactory family without endangering biodiversity (Figure 2.4) (Burger et al., 2017; De Silva, 1995; Sharmeen et al., 2021).

2.5 SCIENTIFIC TECHNIQUES AND THEIR PARALLEL JOURNEY WITH THE POPULARIZATION OF ESSENTIAL OIL

The following methods are mostly employed in research labs to capture trace amounts of volatiles from aromatic plants, and they are also used in part to determine the quantity of EOs in plant material. The most often used apparatus is the circulatory distillation apparatus, which was developed based on

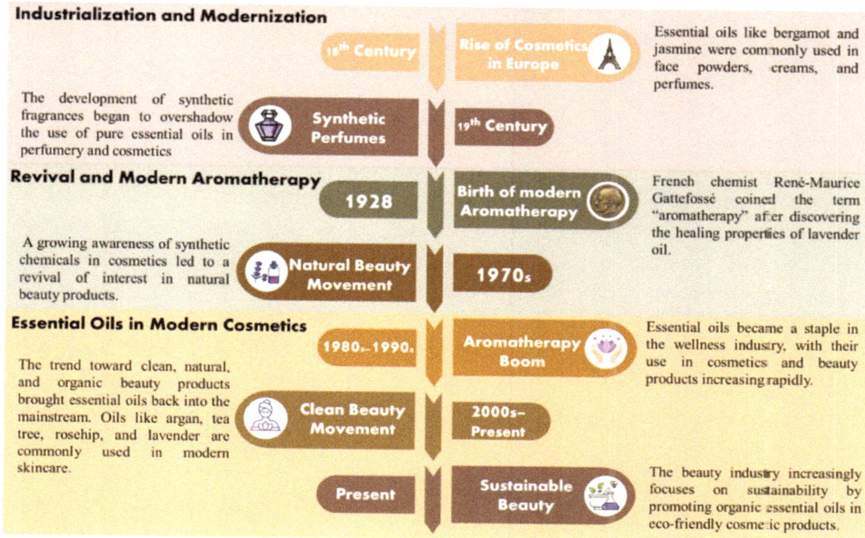

Figure 2.4 Modern change in the use of essential oil and natural compounds driven toward sustainable cosmetics.

Clevenger's 1928 article, *Apparatus for the determination of volatile oil.* and has undergone several revisions since then. A modified version of the apparatus that Cocking and Middleton (1935) described is included in the European Pharmacopoeia as well as in a number of other pharmacopeias. Jennings (1979) published a method for preparing samples based on Soxhlet extraction in a pressurized container with liquid carbon dioxide as the extractant. This apparatus generates extracts without the need of solvents, making them ideal for use in high-resolution gas chromatography (HRGC). Several researchers have suggested the use of microwave-assisted extraction as a less time-consuming option. Craveiro et al. (1989), for example, suggested this method utilizing a round-bottom flask holding fresh plant material. In the years that followed, further chromatographic separation techniques were developed, including thin-layer chromatography, gas chromatography, liquid column chromatography, fluid chromatography, and counter current chromatography. Further advancements in hyphenated techniques have led to the development of a variety of applications, such as gas chromatography-mass spectrometry, high-performance liquid chromatography-gas chromatography, supercritical fluid extraction-gas chromatography, gas chromatography-ultraviolet spectroscopy, gas chromatography-atomic emission spectroscopy, and gas chromatography-Fourier transform infrared spectroscopy. Later, with the advancement in identification of multicomponent samples without previous separation via UV spectroscopy, and IR spectroscopy, mass spectrometry, and 13C-NMR spectroscopy, the use of EOs in different fields was revolutionized (Kubeczka, 2020).

2.6 REGULATION OF USE IN COSMETICS AND ITS EFFECT ON RECENT USE

Aromatic herbs contain EOs, albeit in diluted form. EOs are extremely powerful chemicals that become 100 times more concentrated following distillation from plant components. Because EOs are not subject to the same rigorous testing as pharmaceuticals and are governed by the US Food and Drug Administration (FDA) as cosmetics, it is important to utilize them carefully and educate consumers about them. Despite coming from natural sources, these oils have the potential to cause side effects such as phototoxicity, oral toxicity, allergic and primary contact dermatitis, and ocular irritation. Topical application increases the risk of allergic reactions, local irritation, and photosensitization. It also increases the risk of redness, lacrimation, and respiratory irritation or distress if it accidentally gets in the eyes or nose. Patch testing and dilution are necessary to reduce dangers associated with direct skin contact, particularly with undiluted oils that can cause irritating, worsening rashes. The FDA regulates EOs according to their intended application, which may include cosmetics, drugs, home items, or other uses. It determines a product's intended use based on several factors, including what customers expect from it and the promises made on labels, websites, and marketing (Vora et al., 2024).

If a product—such as cologne, shower gel, perfume, etc. is made only to improve someone's appearance or clean their body, it is considered cosmetic. Laws do not require FDA approval for cosmetics to be sold. The Food and Drugs Act and the Natural Health Product Regulations (the Regulations) by the Natural and Nonprescription Health Product Directorate (NNHPD) serve as Canada's regulatory bodies for EOs. The European Medicine Agency is responsible for providing regulatory rules for EOs in the European Union through Directives 2001/83/EC and 2004/24/EC. The Department of AYUSH in India establishes the recommendations regarding EOs through the Drugs and Cosmetics Act (D and C) of 1940 and Rules of 1945 (Osaili et al., 2023). To sum up, the regulations about EOs guarantee that customers can obtain secure and efficient items (Vora et al., 2024).

2.7 SAFETY CONCERNS ASSOCIATED WITH THE USE OF ESSENTIAL OIL AND ITS YES-NO RELATIONSHIP WITH COSMETICS OVER TIME

Concerns over the toxicity of natural substances in personal care products have surfaced in recent years. Claims that a product is free of parabens, sulfates, PEGs, silicones, phthalates, and an increasing number of other chemicals that have been banned from use in cosmetics can be seen on the labels of many modern products. However, the possible toxicity of natural substances has received far less consideration. The fact that personal care product labeling laws do not ensure that a customer would be informed of potentially dangerous substances is a serious cause for concern. For instance, limonene, citral, citronellol, linalool, and other aroma allergens are present in a lot of plant oils. Under INCI (International Nomenclature of Cosmetic Ingredients) regulations, the product label should list only the plant extract's Latin name, without specifying individual compounds that might pose a risk. Furthermore, there is a worry that the extraction and use of natural hazardous components in personal care products may contaminate our rivers and expose people in the process (Burnett et al., 2019). For instance, it has been demonstrated that isoflavone (phytoestrogen) production in a grassland area can increase to as high as 220 kg per hectare annually. There's a chance that the growing quantity and diversity of plants being harvested for cosmetics will lead to higher concentrations of some naturally occurring hazardous substances in our soil and waterways. Furthermore, cosmetic compounds that wash off our skin and hair when we bathe, or shower may also end up in our waterways.

Since EOs come from nature, they are typically regarded as safe. The natural state of items derived from plants does not negate their potential for toxicity or allergy. While the majority of EOs are thought to be safe, some have the potential to be irritating, sensitizing, phototoxic, or to trigger allergic reactions, including anaphylaxis (Antignac et al., 2011). EOs are classified as either low or medium toxic. In vitro assays using human fibroblasts or dermal epithelial cells, EOs had a cytotoxic impact with CC50 values ranging from 5.0 to 1.950 μg/ml. It is noteworthy that the antibacterial properties of EOs are attained at concentrations of around 20–20,000 μg/ml, far surpassing the range of cytotoxicity. As a result, harmful effects could occur at doses lacking in antibacterial action (Reichling et al., 2009). When using cosmetics containing EOs, skin irritation and sensitization are the most frequent unpleasant reactions. Because of the complexity and diversity of ingredients in oils, determining the sensitizing agent can be challenging. Depending on chemotypes, environment, season, and harvest time, the content of oil can vary greatly, resulting in the synthesis of distinct metabolite and, consequently, varying toxicity (Vigan, 2010). Higher oil concentrations have a greater potential to cause allergies, and sensitization reactions are dose-dependent (Dreger and Wielgus, 2013.

After being exposed to an allergen repeatedly, an allergic reaction occurs. The oil content and species affect the threshold of irritation. The Cosmetic Regulation No. 1223/2009, which went into effect on July 11, 2013, defines the laws of the European Union pertaining to cosmetic product. The Regulation specifies in great depth what needs to be included in product descriptions, safety reports, manufacturing and storage procedures, GMP (good manufacturing practice) compliance reports, ingredient restrictions, pack labeling, and notification. Ingredients Restrictions in Annex II–VI contain a list of compounds that are either prohibited or restricted from being used in cosmetic products, as well as those that are allowed, like preservatives. Despite having antibacterial properties, EOs are not listed as preservatives under the Cosmetic Regulation. They are not included in Annex VI and are acknowledged by the European Scientific Committee as additional elements. We may conclude that EOs are typically safe chemicals based on their antibacterial activity and safety. However, it is important to keep in mind that EOs are complex mixtures of elements, and some of them may be poisonous or dangerous (Regulation (EC) No. 1223/2009; Dreger and Wielgus, 2013).

Moreover, the International Organization for Standardization (ISO) assigns a unique identifier to each EO to codify its composition. Together with the criteria for labeling, transport, and marking, ISO has also standardized several analytical techniques for ensuring the quality control of these specifications in the ISO/Technical Committee. To raise the safety level of these oils and protect consumer health when purchasing products containing EOs, ISO/TC 54 places a strong emphasis on international trade in EOs (Es et al., 2017). In order to provide safe cosmetic goods, the International Fragrance Association (IFRA) also establishes the maximum concentration of certain EOs and components that may pose an allergic risk. Additionally, the IFRA provides guidelines for the safe application of fragrance components, which are published in the IFRA Code of Practice and its guidelines (SCCNFP, 1999).

Depending on their concentration, certain EOs are recognized to cause skin irritation. NAHA states that they should always be correctly diluted in vegetable oil and should not be applied to any inflammatory or allergic skin conditions. Bay (*Pimento racemosa*), cinnamon (*Cinnamomum zeylanicum*), clove bud (*Syzygium aromaticum*), cumin (*Cuminum cyminum*), lemon verbena (*Lippia citriodora*), oregano (*Origanum vulgare*), tagetes (*Tagetes minuta*), and thyme chemotype thymol (*Thymus vulgaris*) are among the plants that cause skin irritation. Additionally, consumers need to be advised that certain EOs may increase their sensitivity to UV radiation and the sun. For at least 24 hours after using photosensitizing EOs, one should stay out of the sun and artificial tanning. According to NAHA, photosensitizing oils include Angelica root (*Angelica archangelica*), bergamot (*Citrus bergamia*), cumin (*Cuminum cyminum*), expressed lemon (*Citrus limon*), expressed lime (*Citrus medica*), expressed bitter orange (*Citrus aurantium*), and rue (*Ruta graveolens*) (NAHA).

2.8 ROLE OF NANOTECHNOLOGY IN MODERN DAY COSMETICS

EO encapsulation has been widely used as a nanotechnology tool to shield these resources from the outside world and adjust their release to suit particular requirements. The encapsulation of bioactive substances is a practical and effective substitute. This method can also make chemicals more physically stable, shield them from environmental interactions, reduce volatility, boost bioactivity, lessen toxicity, and even permit the gradual release of resources

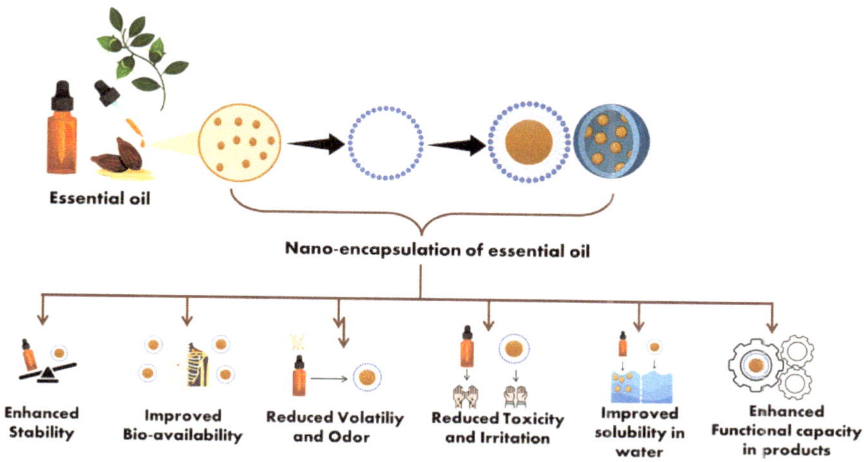

Figure 2.5 Nanotechnology and its role in combating the existing issues related to the use of essential oils in cosmetics.

in certain media (Figure 2.5). The assets included in the nanoparticles (NPs) are displaced to the application medium by the controlled-release mechanism. This displacement is slow and relates to the concentration that is released gradually, providing advantages for the ecosystem and human health such as lessening the evaporation of volatile assets, ease of handling, and a decrease in environmental contaminants and phytotoxicity (Albuquerque et al., 2022; Kujur et al., 2020). Furthermore, cosmeceuticals' look, coverage, and skin adherence are all improved when NPs are added to cosmetic formulations without affecting the products' original qualities. Nanosized chemicals are used by cosmetic producers to enhance a range of features, including UV protection, skin penetration, color, scent release, finish quality, and anti-aging effects. By regulating the transport of active components, producing site-specificity, boosting drug-loading capacity, or improving biocompatibility, they extend the duration of action (Gupta et al., 2022).

2.9 CONCLUSION

As part of a healthy lifestyle, we use a variety of cosmetic items daily for varied purposes. As per the Federal Food, Drug & Cosmetic Act (FD&C), cosmetics are defined as "Articles intended to be rubbed, poured, sprinkled, or sprayed on, introduced into, or otherwise applied to the human body or any part thereof for cleansing, beautifying, promoting attractiveness, or altering the appearance." They not only make us look better, but they also shield our skin and hair from stress and pollution, preserving human health and wellness. In this context, EOs could be used as invaluable resources for the cosmetics industry because of its pleasant aroma and bioactive ingredients. In view of increasing demands of customers for safer and effective cosmetics, an intelligent combination of EOs with available cosmetic products could enhance the industrial application of natural products in the fragrance and aroma industries, especially in the development of cosmetic products in near future. With the growing grasp of social media over almost all age groups of population, it will be interesting to see how the EOs-based products expand themselves in the cosmetic industry in upcoming years.

ACKNOWLEDGMENTS

The authors acknowledge the financial support by the Institute of Eminence (No. R/Dev/D/IOE/Incentive/2021-22/32393), Banaras Hindu University, Varanasi, India.

REFERENCES

Adlard, E. R. (2010). *Handbook of essential oils. Science, technology and applications*, K. Hüsnü Can Başer and Gerhard Buchbauer (Eds.). CRC Press.

Alam, P., Imran, M., Ali, A., &Majid, H. (2024). *Cananga odorata* (ylang-ylang) essential oil containing nanoemul gel for the topical treatment of scalp psoriasis and dandruff. *Gels*, *10*(5), 303.

Albuquerque, P. M., Azevedo, S. G., de Andrade, C. P., D'Ambros, N. C. D. S., Pérez, M. T. M., & Manzato, L. (2022). Biotechnological applications of nanoencapsulated essential oils: A review. *Polymers*, *14*(24), 5495.

Al-Qahtani, W. H., Dinakarkumar, Y., Arokiyaraj, S., Saravanakumar, V., Rajabathar, J. R., Arjun, K., & Appaturi, J. N. (2022). Phyto-chemical and biological activity of *Myristica fragrans*, an Ayurvedic medicinal plant in Southern India and its ingredient analysis. *Saudi Journal of Biological Sciences*, *29*(5), 3815–3821.

Antignac, E., Nohynek, G. J., Re, T., Clouzeau, J., & Toutain, H. (2011). Safety of botanical ingredients in personal care products/cosmetics. *Food and Chemical Toxicology*, *49*(2), 324–341.

Ao, Y., Satoh, K., Shibano, K., Kawahito, Y., & Shioda, S. (2008). Singlet oxygen scavenging activity and cytotoxicity of essential oils from rutaceae. *Journal of Clinical Biochemistry and Nutrition*, *43*(1), 6–12.

Apone, F., Barbulova, A., & Colucci, M. G. (2019). Plant and microalgae derived peptides are advantageously employed as bioactive compounds in cosmetics. *Frontiers in Plant Science*, *10*, 756.

Baser, K. H. C., & Buchbauer, G. (2009). *Handbook of essential oils: Science, technology, and applications*. CRC Press.

Bousbia, N., Vian, M. A., Ferhat, M. A., Petitcolas, E., Meklati, B. Y., & Chemat, F. (2009). Comparison of two isolation methods for essential oil from rosemary leaves: Hydrodistillation and microwave hydrodiffusion and gravity. *Food Chemistry*, *114*(1), 355–362.

Burger, P., Landreau, A., Watson, M., Janci, L., Cassisa, V., Kempf, M., & Fernandez, X. (2017). Vetiver essential oil in cosmetics: What is new? *Medicines*, *4*(2), 41.

Burger, P., Plainfossé, H., Brochet, X., Chemat, F., & Fernandez, X. (2019). Extraction of natural fragrance ingredients: History overview and future trends. *Chemistry & Biodiversity*, *16*(10), e1900424.

Burnett, C. L., Fiume, M. M., Bergfeld, W. F., Belsito, D. V., Hill, R. A., Klaassen, C. D., & Heldreth, B. (2019). Safety assessment of citrus-derived peel oils as used in cosmetics. *International Journal of Toxicology*, *38*(2_suppl), 33S–59S.

Butnariu, M., & Sarac, I. (2018). Essential oils from plants. *Journal of Biotechnology and Biomedical Science*, 1(4), 35–43.

Çeçen, F., & Tezel, U. (Eds.). (2017). *Hazardous pollutants in biological treatment systems: Fundamentals and a guide to experimental research.* IWA Publishing.

Cholke, P. B., Bhor, A. K., Shete, A. M., Sonawane, R. K., & Pravin, C. (2017). Extraction and GC-MS analysis of orange (*Citrus sinensis*) peel oil. *Research Journal of Life Sciences, Bioinformatics, Pharmaceutical and Chemical Sciences*, *2*(5), 41–51.

Clevenger, J. F. (1928). Apparatus for the determination of volatile oil. *The Journal of the American Pharmaceutical Association (1912)*, *17*(4), 345–349.

Cocking, T. T., & Middleton, G. (1935). *An improved method for the estimation of the essential oil content of drugs.* Pharmaceutical Press.

Craveiro, A. A., Matos, F. J. A., Alencar, J. W., & Plumel, M. M. (1989). Microwave oven extraction of an essential oil. *Flavour and Fragrance Journal*, *4*(1), 43–44.

De Silva, T. U. L. E. Y. (1995). Development of essential oil industries in developing countries. K. Tuley de silva, Nov, 1.

Dini, I., & Laneri, S. (2021). The new challenge of green cosmetics: Natural food ingredients for cosmetic formulations. *Molecules*, *26*(13), 3921.

Donelian, A., Carlson, L. H. C., Lopes, T. J., & Machado, R. A. F. (2009). Comparison of extraction of patchouli (*Pogostemon cablin*) essential oil with supercritical CO_2 and by steam distillation. *The Journal of Supercritical Fluids*, 48(1), 15–20.

Dreger, M., & Wielgus, K. (2013). Application of essential oils as natural cosmetic preservatives. *Herba polonica*, *59*(4).

Es, I., Khaneghah, A. M., & Akbariirad, H. (2017). Global regulation of essential oils. In M. Shahidi & A. M. Khaneghah (Eds.), *Essential Oils in Food Processing: Chemistry, Safety and Applications* (pp. 327–338). Hoboken, NJ: John Wiley & Sons.

Ezekwe, N., King, M., & Hollinger, J. C. (2020). The use of natural ingredients in the treatment of alopecias with an emphasis on central centrifugal *Cicatricial alopecia*: A systematic review. *The Journal of Clinical and Aesthetic Dermatology*, *13*(8), 23.

Figueiredo, A. C., Barroso, J. G., Pedro, L. G., & Scheffer, J. J. (2008). Factors affecting secondary metabolite production in plants: Volatile components and essential oils. *Flavour and Fragrance Journal*, *23*(4), 213–226.

Ghazghazi, H., Miguel, M. G., Weslati, M., Hasnaoui, B., Sebei, H., Barroso, J. G., & Figueiredo, A. C. (2012). Chemical variability of the essential oils from *Rosa canina* L. and *Rosa sempervirens* L. flowers collected at Tunisia. *Journal of Essential Oil Research*, *24*(5), 475–480.

Ghorbani, A., & Esmaeilizadeh, M. (2017). Pharmacological properties of Salvia officinalis and its components. *Journal of Traditional and Complementary Medicine*, *7*(4), 433–440.

Giang, P. M., & Son, P. T. (2016). GC and GC-MS analysis of the fresh flower essential oil of *Cananga odorata* (lam.) Hook. f. et Th. var. fruticosa (Craib) J. Sincl. *American Journal of Essential Oils and Natural Products*, *4*(4), 09–11.

Gilani, A. H., Khan, A. U., Jabeen, Q., Subhan, F., & Ghafar, R. (2005). Antispasmodic and blood pressure lowering effects of *Valeriana wallichii* are mediated through K+ channel activation. *Journal of Ethnopharmacology*, 100(3), 347–352.

Gupta, V., Mohapatra, S., Mishra, H., Farooq, U., Kumar, K., Ansari, M. J., & Iqbal, Z. (2022). Nanotechnology in cosmetics and cosmeceuticals—a review of latest advancements. *Gels*, *8*(3), 173.

Guzmán, E., & Lucia, A. (2021). Essential oils and their individual components in cosmetic products. *Cosmetics*, *8*(4), 114.

Heuberger, E., Hongratanaworakit, T., Böhm, C., Weber, R., & Buchbauer, G. (2001). Effects of chiral fragrances on human autonomic nervous system parameters and self-evaluation. *Chemical Senses*, 26(3), 281–292.

Huang, X. W., Feng, Y. C., Huang, Y., & Li, H. L. (2013). Chemical composition, antioxidant and the possible use as skin-care ingredient of clove oil (*Syzygium aromaticum* (L.) Merr. & Perry) and citronella oil (*Cymbopogon goeringii*) from China. *Journal of Essential Oil Research*, *25*(4), 315–323.

Hussain, A. I., Anwar, F., Chatha, S. A. S., Jabbar, A., Mahboob, S., & Nigam, P. S. (2010). *Rosmarinus officinalis* essential oil: Antiproliferative, antioxidant and antibacterial activities. *Brazilian Journal of Microbiology*, *41*, 1070–1078.

Hussain, A. I., Anwar, F., Sherazi, S. T. H., &Przybylski, R. (2008). Chemical composition, antioxidant and antimicrobial activities of basil (*Ocimum basilicum*) essential oils depends on seasonal variations. *Food Chemistry*, *108*(3), 986–995.

Irshad, M., Subhani, M. A., Ali, S., & Hussain, A. (2020). Biological importance of essential oils. *Essential Oils-Oils of Nature*, *1*, 37–40.

Jennings, W. G. (1979). Vapor-phase sampling. *Journal of High Resolution Chromatography*, 2(5), 221–224.

Kozuharova, E., Simeonov, V., Batovska, D., Stoycheva, C., Valchev, H., & Benbassat, N. (2023). Chemical composition and comparative analysis of lavender essential oil samples from Bulgaria in relation to the pharmacological effects. *Pharmacia*, *70*, 395–403.

Kubeczka, K. H. (2020). History and sources of essential oil research. In K. H. C. Başer & G. Buchbauer (Eds.), *Handbook of Essential Oils: Science, Technology, and Applications* (pp. 3–39). CRC Press: Boca Raton, FL.

Kujur, A., Kumar, A., Yadav, A., & Prakash, B. (2020). Antifungal and aflatoxin B1 inhibitory efficacy of nanoencapsulated *Pelargonium graveolens* L. essential oil and its mode of action. *LWT* 130, 109619.

Kulisic, T., Radonic, A., Katalinic, V., & Milos, M. (2004). Use of different methods for testing antioxidative activity of oregano essential oil. *Food Chemistry*, 85(4), 633–640.

Kusuma, H. S., & Mahfud, M. (2017). GC-MS analysis of essential oil of Pogostemon cablin growing in Indonesia extracted by microwave-assisted hydrodistillation. *International Food Research Journal*, *24*(4), 1525.

Lahlou, M. (2004). Methods to study the phytochemistry and bioactivity of essential oils. *Phytotherapy Research: An International Journal Devoted to Pharmacological and Toxicological Evaluation of Natural Product Derivatives*, *18*(6), 435–448.

Laska, M., & Teubner, P. (1999). Olfactory discrimination ability of human subjects for ten pairs of enantiomers. *Chemical Senses*, *24*(2), 161–170.

Lucia, A., Toloza, A. C., Guzmán, E., Ortega, F., & Rubio, R. G. (2017). Novel polymeric micelles for insect pest control: Encapsulation of essential oil monoterpenes inside a triblock copolymer shell for head lice control. *PeerJ*, *5*, e3171.

McMullen, R. L. (2018). *Antioxidants and the skin*. CRC Press.

McMullen, R. L., & Dell'Acqua, G. (2023). History of natural ingredients in cosmetics. *Cosmetics*, 10(3), 71.

Morocho-Jácome, A. L., Freire, T. B., de Oliveira, A. C., de Almeida, T. S., Rosado, C., Velasco, M. V. R., & Baby, A. R. (2021). In vivo SPF from multifunctional sunscreen systems developed with natural compounds—A review. *Journal of Cosmetic Dermatology*, *20*(3), 729–737.

Moy, R. L., & Levenson, C. (2017). Sandalwood album oil as a botanical therapeutic in dermatology. *The Journal of Clinical and Aesthetic Dermatology*, *10*(10), 34.

Muggli, R. (2005). Systemic evening primrose oil improves the biophysical skin parameters of healthy adults. *International Journal of Cosmetic Science*, *27*(4), 243–249.

Naeem, A., Abbas, T., Ali, T. M., & Hasnain, A. (2018). Essential oils: Brief background and uses. *Annals of Short Reports*, *1*(1), Article 1006.

Navarra, M., Mannucci, C., Delbò, M., & Calapai, G. (2015). *Citrus bergamia* essential oil: From basic research to clinical application. *Frontiers in Pharmacology*, 6, 36.

Olujimi, O. O., Fatoki, O. S., Odendaal, J. P., & Okonkwo, J. O. (2010). Endocrine disrupting chemicals (phenol and phthalates) in the South African environment: A need for more monitoring. *Water SA*, 36(5), 671–682.

Osaili, T. M., Dhanasekaran, D. K., Zeb, F., Faris, M. E., Naja, F., Radwan, H., & Obaid, R. S. (2023). A status review on health-promoting properties and global regulation of essential oils. *Molecules*, *28*(4), 1809.

Parham, S., Kharazi, A. Z., Bakhsheshi-Rad, H. R., Nur, H., Ismail, A. F., Sharif, S., & Berto, F. (2020). Antioxidant, antimicrobial and antiviral properties of herbal materials. *Antioxidants*, *9*(12), 1309.

Piesse, G. W. S. (1879). *The art of perfumery and the methods of obtaining the odours of plants*. Longmans.

Phetcharat, L., Wongsuphasawat, K., & Winther, K. (2015). The effectiveness of a standardized rose hip powder, containing seeds and shells of *Rosa canina*, on cell longevity, skin wrinkles, moisture, and elasticity. *Clinical Interventions in Aging*, 10, 1849–1856.

Prusinowska, R., & Smigielski, K. B. (2014). Composition, biological properties and therapeutic effects of lavender (Lavandula angustifolia L.). A review. *Herba polonica*, 60(2).

Pujari, R., Reddy, S. B., & Satishkumar, M. (2020). Essential oils in India-history and future prospects. *Biotica Research Today*, *2*(9), 938–940.

Rajeswara, B. R., & Bhattacharya, A. K. (1992). History and botanical nomenclature of rose scented geranium cultivars grown in India. *Indian Perfumer*, *36*, 155.

Raghuvanshi, T. S., Singh, P. P., Kohar, N., & Prakash, B. (2023). Essential oils: From traditional to modern-day applications with special reference to medicinal and aromatic plants in India. In T. S. Raghuvanshi, P. P. Singh, N. Kohar, & B. Prakash (Eds.), *Plant essential oils: From traditional to modern-day application* (pp. 1–26). Singapore: Springer Nature Singapore

Rahmi, D., Yunilawati, R., Jati, B. N., Setiawati, I., Riyanto, A., Batubara, I., & Astuti, R. I. (2021). Antiaging and skin irritation potential of four main Indonesian essential oils. *Cosmetics*, *8*(4), 94.

Reichling, J., Schnitzler, P., Suschke, U., & Saller, R. (2009). Essential oils of aromatic plants with antibacterial, antifungal, antiviral, and cytotoxic properties–an overview. *Forschende Komplementärmedizin/Research in Complementary Medicine*, *16*(2), 79–90.

Sarkic, A., &Stappen, I. (2018). Essential oils and their single compounds in cosmetics—A critical review. *Cosmetics*, *5*(1), 11.

Sawamura, M. (Ed.). (2011). *Citrus essential oils: Flavor and fragrance*. John Wiley & Sons.

SCCNFP. (1999). The Scientific Committee on Cosmetic Products and Non-Food Products Intended for Consumers, Fragrance Allergy in Consumers. A Review of the Problem. Analysis of the Need for Appropriate Consumer Information and Identification of Consumer Molecules 2021, 26, 666 24 of 24 Allergens. Available online: https://ec.europa.eu/health/ph_risk/committees/sccp/documents/out98_en.pdf (accessed on 22 October, 2024).

Sell, C. S., & Pybus, D. H. (2006). *The history of aroma chemistry and perfume.* In *D. H. Pybus & C. S. Sell (Eds.), The Chemistry of Fragrances: From Perfumer to Consumer* (pp. 3–23). Royal Society of Chemistry.

Sharmeen, J. B., Mahomoodally, F. M., Zengin, G., & Maggi, F. (2021). Essential oils as natural sources of fragrance compounds for cosmetics and cosmeceuticals. *Molecules*, 26(3), 666.

Shivanand, P., Nilam, M., & Viral, D. (2010). Herbs play an important role in the field of cosmetics. *International Journal of PharmTech Research*, 2(1), 632–639.

Sun, P., Zhao, L., Zhang, N., Wang, C., Wu, W., Mehmood, A., & Zhou, F. (2020). Essential oil and juice from bergamot and sweet orange improve acne vulgaris caused by excessive androgen secretion. *Mediators of Inflammation*, 2020(1), 8868107.

Vigan, M. (2010). Essential oils: Renewal of interest and toxicity. *European Journal of Dermatology*, *20*(6), 685–692.

Vora, L. K., Gholap, A. D., Hatvate, N. T., Naren, P., Khan, S., Chavda, V. P., & Khatri, D. K. (2024). Essential oils for clinical aromatherapy: A comprehensive review. *Journal of Ethnopharmacology*, 118180.

Zhao, H., Ren, S., Yang, H., Tang, S., Guo, C., Liu, M., & Xu, H. (2022). Peppermint essential oil: Its phytochemistry, biological activity, pharmacological effect and application. *Biomedicine & Pharmacotherapy*, *154*, 113559.

Chapter 3

Safety and toxicological assessments of essential oils and fragrances in cosmetics

Priyanka Rana, Prakriti Gaur, and Sunil Kumar Patel
CSIR-Indian Institute of Toxicology Research, Lucknow, India
and
Academy of Scientific and Innovative Research (AcSIR), Ghaziabad, India

Medha Bajpai
CSIR-Indian Institute of Toxicology Research, Lucknow, India

Ayushi Mishra
Department of Biochemistry, University of Lucknow, Lucknow, India

3.1 INTRODUCTION

Fragrances or essential oil, primarily incorporated into cosmetics, personal care products and household goods, are to hide undesirable odors or to provide aesthetically attractive aromas. These fragrances consist of mixtures of volatile organic molecules that emit unique and pleasing scents. In cosmetics formulations, they play a crucial role in cosmetic products by masking unpleasant odors and enhancing the overall sensory appeal of the product. However, many chemical substances of fragrances to which humans exposed through multiples routes including skin, oral, and via inhalation [1]. Exposure to fragrance ingredients could induce adverse health effects, highlighting the need to ensure the consumer's safety. The presence of fragrances in perfumes and other personal care products has been linked to a range of negative consequences and majorly, these fragrance compounds are responsible for allergic reactions such as contact dermatitis. Therefore, it is essential to evaluate the sensitization potential of these compounds, and in this context, this, chapter will review the risk assessment approaches used to determine their sensitizing potency. The risk assessment of fragrance ingredients is a structured scientific method aimed to provide cosmetics consumers with reliable safety information and estimating the probability of toxic effects induced by fragrance or essential oil ingredients. Traditionally, fragrance risk assessment has been performed in three steps: (i) an exhaustive survey of published literature, (ii) evaluation of read across materials, and (iii) dermal sensitization threshold (DST) determination. Based on these

DOI: 10.1201/9781003558477-3

evaluations, no expected sensitization induction level (NESIL) is established, through which quantitative risk assessment (QRA) is calculated and sensitizing potency is determined. Notably, since 2013, the European Union (EU) has prohibited animal-based toxicity testing of cosmetic products and their components, necessitating the adoption of alternative non-animal testing methods in fragrances and their ingredients safety evaluations [2]. For many years, toxicologists have identified the development of non-animal alternatives as a top priority for chemical safety evaluation. In this chapter, we present and discuss the adverse outcome pathway (AOP) framework that covers toxicological processes testing for evaluating the skin sensitization. The skin sensitization AOP describes the biological processes leading to allergic contact dermatitis (ACD) through four defined key events (KEs). Several *in chemico* and *in vitro* methods developed based on these key events, which have been validated and adopted as a test guideline by the Organization for Economic Co-operation and Development (OECD). Specially, the direct peptide reactivity (DPRA) assay, assesses the haptenation or chemical reactivity with skin proteins corresponding as key event 1. The KeratinoSens™ assay measures the keratinocyte activation representing the key event 2. The human cell line activation test (h-CLAT) and U937 cell line activation test evaluate the dendritic cell activation corresponding to key event 3. Finally, the local lymph node assay (LLNA) assess T-cell proliferation, traditionally considered as the key event 4 [3]. In addition, allergen labeling is an important tool to reduce the risk of exposure and prevent sensitization in peoples with allergic contact dermatitis. This chapter outlines the current status of allergen labeling regulations around the world and critically, reviews the value of current existing precautionary recommendations intended to protect the allergic consumers.

3.2 RISK ASSOCIATED WITH FRAGRANCES

Numerous fragranced products including perfumes, colognes, body soaps, hand washes, shampoos, facial/hand cleansers, creams, deodorants, as well as household items such as air fresheners, disinfectants, fragrance diffusers, laundry and dish detergents, scented candles, surface cleaners are utilized daily on a global scale. Many of these scented items contain low-molecular-weight compounds that may induce skin sensitization and contribute to development of ACD upon repeated exposure [4]. ACD is an inflammatory skin disease characterized by a type 4 hypersensitivity response, which occurs when an irritating chemical or a sensitizing antigen comes into contact with the skin, penetrates it, and subsequently triggers T-cell-mediated immune reaction. The frequency of contact dermatitis believed to be between 1.7% and 9.8%, [5]. So, to safeguard consumers, all fragrance or essential oil containing products must undergo a rigorous evaluation, including safety assessments and toxicological testing, prior to their introduction to the market.

3.3 RISK ASSESSMENT METHODOLOGIES

Risk assessment methodologies for ACD provide a structured framework to evaluate the sensitization potential of chemicals and ensure consumer safety. These methodologies combine evidence from toxicological data, non-animal test methods, threshold-based evaluations, and quantitative risk assessment (QRA) approaches to determine the potency and safe exposure levels of sensitizers.

Step 1: Look at all the written and unpublished materials thoroughly to establish the current state of knowledge about a certain fragrance ingredient. This involves examining the safety of fragrance materials, with available data for relevant toxicological endpoints for both local and systemic effects. If enough accurate information is available about the endpoint, a risk assessment can be conducted based on an endpoint evaluation; if the data are inadequate or of insufficient quality, advance to Step 2.

Step 2: Review metabolism, chemistry, physico-chemical characteristics, and toxicological features for read-across to analogs found and considered suitable for each endpoint. Read-across method uses common understanding of a certain characteristic, such as physico-chemical properties and toxicity profiles of one or more chemicals to predict the same characteristic for another target molecule. This technique may assist in circumventing the need of conducting individual tests on each chemical for each toxicological endpoint. If the substances lacking an appropriate read-across and for which no NESIL be established are forwarded to Step 3.

Step 3: This step is assessed using the DST, which utilizes the principle of the threshold of toxicological concern (TTC) for assessing dermal sensitization. The DST defines a threshold below which the risk of sensitization is negligible, relying on a probabilistic analysis of potency data from various known chemical allergens. For the non-reactive chemicals, a DST of 900 μg/cm^2 is assigned, whereas for reactive substances a lower DST of 64 μg/cm^2 is designated [6]. The NESIL derived from this evaluation, is subsequently used in a QRA to determine the potency of sensitizer (Figure 3.1).

3.4 TOXICOLOGICAL THRESHOLD CONCERN FOR ESSENTIAL OILS

The integration of the TTC, with the use of fragrances, essential oils, and cosmetic products in dermal applications involves a multifaceted approach to ensure safety, efficacy, and consumer satisfaction. Essential oils are pivotal in cosmetic products for their aromatic properties and potential health benefits. The TTC of essential oils provides a framework for assessing the safety of these ingredients at low exposure levels which provides threshold values to ensure that exposure levels remain below those concentrations that

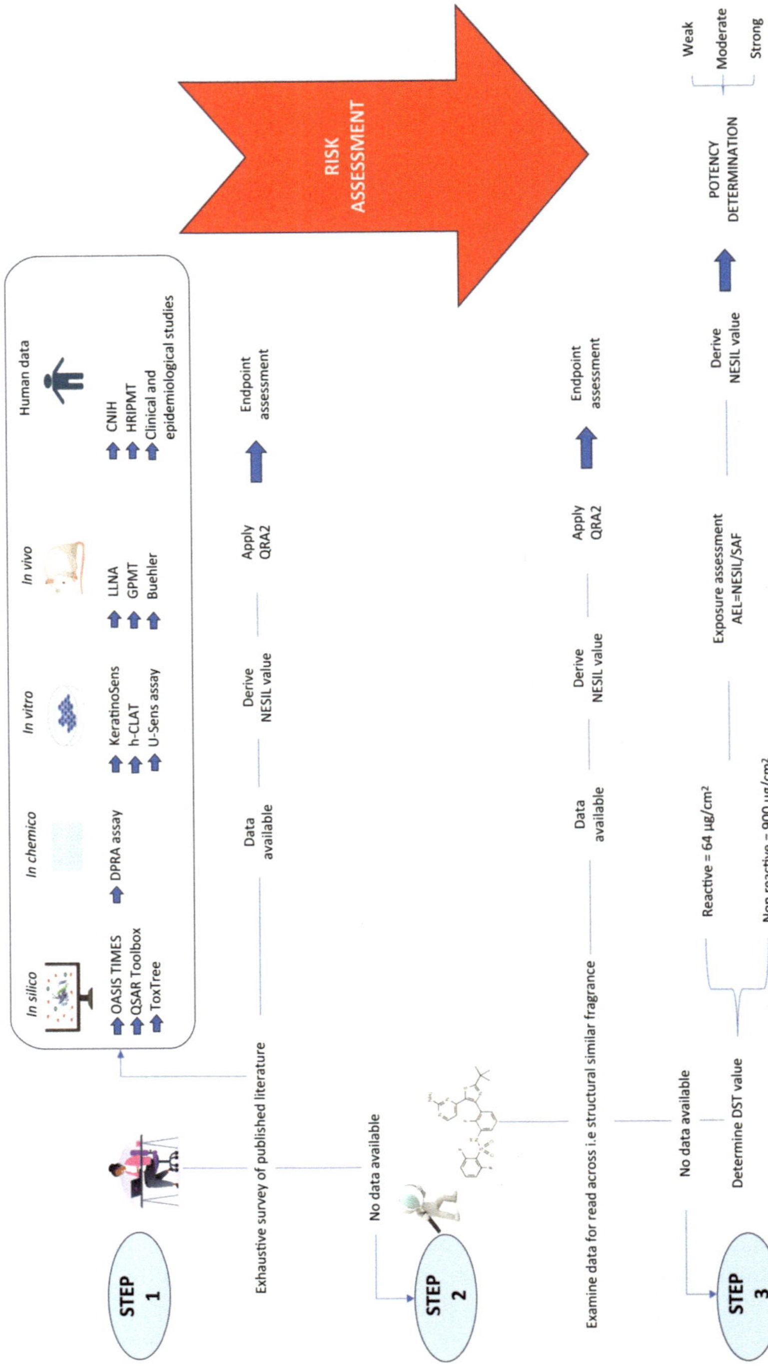

Figure 3.1 Overview of risk assessment methodologies for fragrance materials.

could cause harm. Essential oils are complex mixtures of numerous volatile organic compounds, including terpenes, phenolics, and esters. Some of these components are recognized as safe at low concentrations, others may pose toxicological risks. Evaluating the toxicity of every compound in an essential oil mixture is often time-consuming and resource-intensive. Therefore, the TTC provides a pragmatic framework for evaluating risk based on the chemical structure and known toxicity of related substances. Thus, this framework enables the estimation of safe use levels, minimizes the need for extensive testing, and supports reducing the need for animal testing or exhaustive studies and use of non-animal approaches in fragrance and cosmetic safety evaluation (Figure 3.2).

3.5 TOXICOLOGICAL TESTING REQUIREMENT

An animal testing ban was implemented by the EU in 2009, including both cosmetics goods and their components. Building on this legislation, effective from March 1, 2013, a marketing restriction was introduced under cosmetics regulations (EC) No. 1223/2009, prohibiting the sale of products if the safety testing was conducted on animals after this date, regardless of where the testing took place [7]. In practice, animal testing is now permitted only in rare and extreme cases, such as when no validated alternative methods exist, and where it is necessary to comply with other regulatory frameworks. However, for chemicals registered under the European Chemicals Legislation (Registration, Evaluation, Authorization and Restriction of Chemicals [REACH]), skin sensitization data must still be provided, albeit using validated non-animal methods whenever possible [8]. The OECD released the AOP for skin sensitization in 2012 [9]. The AOP is a conceptual framework that outlines a series of events that occur when an organism or a particular tissue is exposed to a chemical. These events include a number of crucial molecular and cellular processes, ultimately resulting in an unfavorable consequence for the exposed organism or tissue. The skin sensitization AOP was among the first formally established AOPs and maps the mechanistic process leading to ACD in humans. This pathway integrates two major phases:

(i) induction phase, where the immune system is initially sensitized to the chemical allergen, and (ii) elicitation phase, where re-exposure to the allergen triggers an allergic response. Within this framework, the process is summarized into 11 biological stages, of which 4 are recognized as KEs critical to the development of ACD:

KE1 – molecular initiating event; covalent binding of the chemical (hapten) to skin proteins.

KE2 – activation of keratinocytes and release of pro-inflammatory mediators.

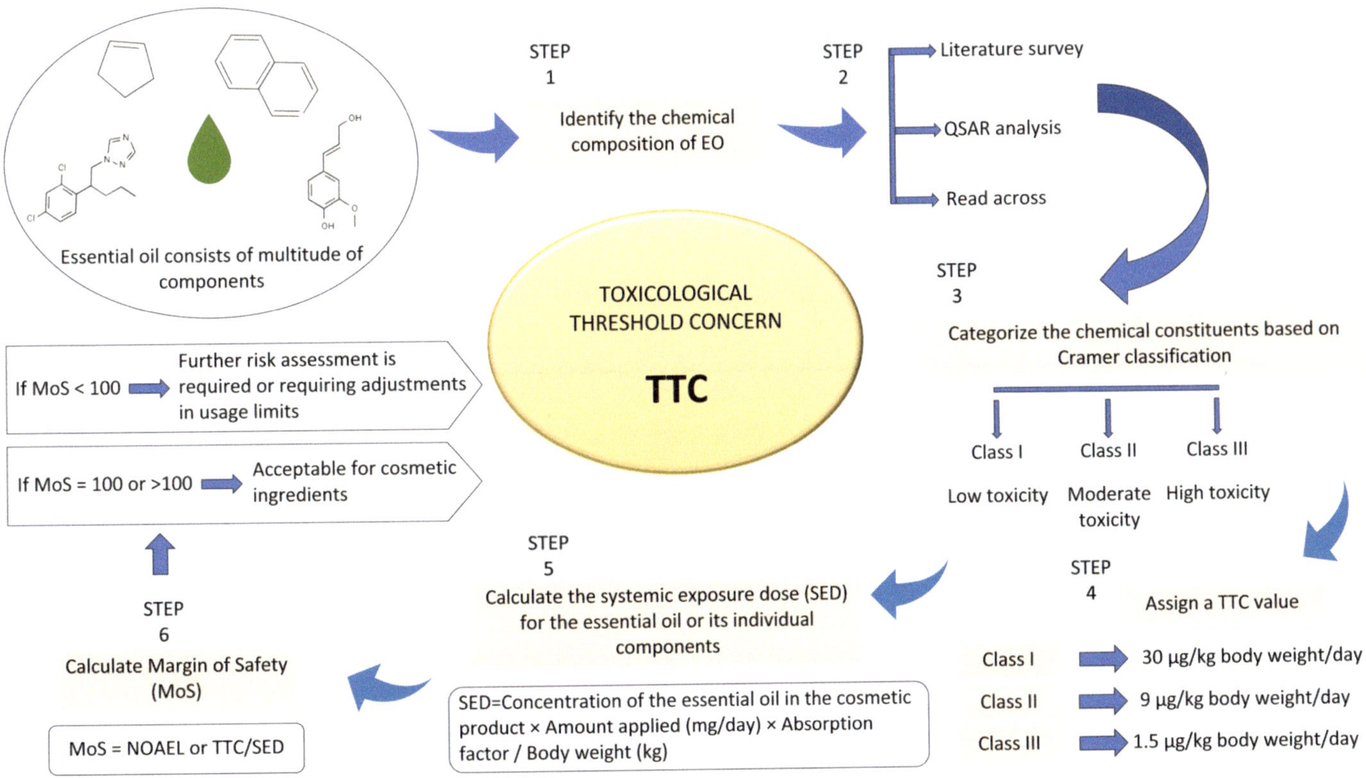

Figure 3.2 Determination of TTC value for the individual constituents of an essential oil.

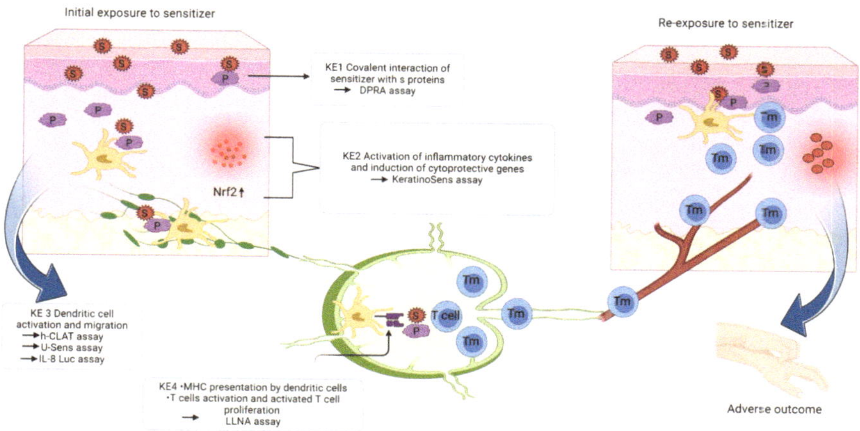

Figure 3.3 A general review of the skin sensitization adverse outcome pathway (AOP).

KE3 – activation and maturation of dendritic cells, including antigen presentation.

KE4 –adverse outcome; activation and proliferation of allergen-specific T-cells, leading to the clinical manifestation of ACD [10] (Figure 3.3).

3.5.1 Key event 1-based method

3.5.1.1 DPRA

The DPRA replicates the phenomenon of haptenation, which refers to the covalent binding of low-molecular weight compounds ("haptens") to proteins. This method assesses the reactivity of chemicals toward synthetic peptides containing either lysine (-AcRFAAKAACOOH) or cysteine (-AcRFAACAACOOH). This approach represents the KE1 in the skin sensitization AOP, also known as molecular initiating event (MIE) [11]. This process is regarded as a fundamental mechanism by which chemicals (or their metabolites) acquire antigenic properties. More precisely, the DPRA measures the reactivity of the test chemical toward synthetic peptide containing either cysteine or lysine amino acids, during a 24-hour incubation period. The test utilizes the rate of depletion of these peptides as the key measurement, which is evaluated using a commonly accessible HPLC-UV technique [12].

3.5.2 Key event 2-based method

3.5.2.1 KeratinoSens™ assay

A cytoprotective pathway activation in keratinocytes is measured using the KeratinoSens™ assay to contribute to the assessment of skin sensitization potential of chemicals. This method addresses the second key event (KE2) of the skin sensitization AOP [13]. The technique operates using a luciferase reporter gene system to identify the activation of the Keap-1/Nrf2 signaling pathway during the covalent interaction between a putative sensitizer and cysteine residues on Keap-1, leading to Nrf2 stabilization and nuclear translocation. In turn, Nrf2 binding to ARE sequences induces transcriptional activation of downstream cytoprotective and detoxification genes, which is captured as a quantifiable luciferase signal that serves as an indicator of keratinocyte activation [14].

3.5.3 Key event 3-based method

3.5.3.1 The human cell line activation test (h-CLAT)

This method quantifies changes in the expression of cell surface markers associated with the process of activation of monocytes and dendritic cells to address the third key event (KE3) of the skin sensitization AOP [15]. The h-CLAT is a surrogate model for dendritic cells activation that quantifies the expression of the co-stimulatory CD86 and CD54 membrane markers in THP-1 cells, a human monocytic leukemia cell line [16]. Flow cytometry is employed to measure the expression levels of markers after 24 hours of exposure to serial concentrations of the test chemical, which are selected based on a pre-determined CV75 (or the concentration of the test chemical that enables 75% of cell survival). A test substance is considered to have a positive dendritic cell-activating potential if it induces a 2-fold increase in CD54 expression and/or a 1.50-fold increase in CD86 expression at a relative cell viability of at least 50% [17]. Upregulation of these markers in THP-1 cells reflects the maturation and activation status of dendritic cells, during skin sensitization.

3.5.3.2 U-SENS™ assay

The U-SENS™ assay monitors the third key event (KE3) of skin sensitization by measuring the alteration in the expression of a cell surface marker (CD86) related to the activation of monocytes and dendritic cells. Specifically, it measures changes in the expression of the co-stimulatory surface marker CD86 in the U937 human histiocytic lymphoma cell line after exposure to potential sensitizers [18]. The quantified expression levels of the CD86 cell surface marker in the U937 cell line are then used to reinforce the differentiation between skin sensitizers and non-sensitizers, thereby strengthen its utility as an alternative, animal-free approach in integrated testing strategies for skin sensitization.

3.5.3.3 Interleukin-8 reporter gene assay

An IL-8 Luc assay measures alterations in IL-8 expression linked to dendritic cell activation. Using a stable THP-1-derived IL-8 reporter cell line, THP-G8, which is defined by the presence of stable luciferase orange (SLO) and stable luciferase red (SLR) genes controlled by IL-8 promoter and glyceraldehyde 3-phosphate dehydrogenase (GAPDH) promoters, respectively. Following exposure to a test chemical, luminescence is measured to quantitatively assess IL-8 promoter activation, which serves as an indicator of pro-inflammatory signaling relevant to skin sensitization. Sensitizers could be predicted with an accuracy of 95.2% and sensitivity of 93.5% by combining CD86 expression with IL-8 secretion [19]. This dual-parameter approach improves reliability by linking dendritic cell activation markers with cytokine-driven inflammatory responses.

3.5.4 Key event 4-based method

3.5.4.1 Local lymph node assay

The local lymph node assay (LLNA) is a murine model that was created to assess the cutaneous sensitization potential of compounds. The LLNA is a viable alternative to conventional guinea pig-based methods and, in contrast, offers substantial animal welfare advantages [20]. The fundamental principle of the LLNA is that sensitizers stimulate the proliferation of lymphocytes in the draining lymph nodes that are located near the site of test substance application. This proliferation is directly proportional to the potency and dose of the allergen that was administered, and offers a straightforward method for quantifying skin sensitization. The mean proliferation of each test group is compared to the mean proliferation of the vehicle-treated control (VC) group in order to quantify lymphocyte proliferation. The response is expressed as the stimulation index (SI), calculated as the ratio of the mean proliferation in each treated group to that in the concurrent VC group. Thus, an SI value of ≥ 3 is considered the threshold for classifying a test substance as a potential skin sensitizer [21].

3.5.5 General safety testing method for fragrances

3.5.5.1 Acute toxicity

The assessment of acute toxicity in cosmetics is a critical component of ensuring consumer safety, especially given the potential presence of harmful substances. Traditional methods of toxicity testing, which often involve animal testing, have been increasingly replaced by alternative methods due to ethical concerns and regulatory changes, particularly in the EU. These alternative methods include *in vitro* and *in silico* approaches, which are gaining traction for their ability to provide reliable safety data without using animals. The transition to these methods is supported by advancements in scientific research and regulatory frameworks in cosmetic safety evaluation.

The integration of organ-on-chip (OoC) models into the risk assessment of cosmetics, particularly for acute toxicity, represents a significant advancement in the next generation toxicological testing. These microfluidic devices simulate human organs structural and functional aspects by incorporating living cells within dynamic, three-dimensional microenvironments, offering a more accurate and ethical alternative to traditional animal testing. OoC technology is particularly promising for evaluating the safety of cosmetic ingredients, as it can mimic human physiological responses more closely than conventional *in vitro* methods. Importantly, the adoption of OoC technology approach aligns with the 3Rs principle (replacement, reduction, and refinement) by reducing reliance on animal models and enhancing the predictability of human toxicity responses [22].

3.5.5.2 Skin irritation

The assessment of dermal irritation in cosmetics is crucial for ensuring consumer safety. Dermal irritation is described as an inflammatory condition of skin (dermatitis) that is not caused by an immune response [23]. The *in vivo* Draize test, traditionally used for assessing dermal irritation, has faced significant criticism due to ethical concerns, variability in outcomes, and its limited applicability to human biology. Consequently, there has been a shift toward developing *in vitro* alternatives that can provide reliable and human-relevant testing methods. These alternatives focus on using reconstructed human epidermis (RHE) models and other innovative techniques to assess skin irritation potential without the use of live animals, in line with regulatory and ethical advancements.

The RHE model, compliant with OECD Test Guideline 439, offers a reliable *in vitro* alternative to traditional animal testing, aligning with the 3Rs principles (replacement, reduction, refinement) and regulatory requirements. This model is used to evaluate the irritation potential of cosmetic products by simulating human skin's barrier function and response to irritants or sensitizers [24].

3.5.5.3 Phototoxicity

Phototoxic reactions, also known as photoirritation, are commonly observed and characterized by an acute (minutes to hours), non-immunologic skin response in areas exposed to UV radiation or sunlight, typically resembling an exaggerated sunburn. The 3T3 neutral red uptake (NRU) phototoxicity test represents a validated *in vitro* approach for evaluating photo-induced irritation. The evaluation of photo-cytotoxicity involves analyzing the comparative decrease in cell viability after exposure to the sample, both in the presence and absence of UV-Vis irradiation. This method, despite its high sensitivity (93%), is unable to predict photogenotoxicity, photomutagenicity, photocarcinogenicity, or photoallergic (photosensitization) responses [25].

To overcome some of these limitations, use of *in vitro* human 3-D epidermis models for phototoxicity testing in cosmetics is a significant advancement in assessing the safety of cosmetic products and their ingredients. These models provide a more accurate representation of human skin compared to traditional 2D models, allowing for better prediction of phototoxic reactions. Among these, the EpiDerm™ H3D–PT assay is intended to assess the phototoxic potential and potency of topically administered chemicals and formulations. By permitting the application of test substances to the air-exposed surface (stratum corneum), the assay simulates *in vivo* skin conditions, and thereby facilitating the prediction of the phototoxic potency of test materials at usage concentrations relevant to cosmetic use [26].

3.5.5.4 Reproductive toxicity/endocrine disruption

Reproductive toxicity refers to the detrimental impacts of a substance on any aspect of the reproductive cycle, encompassing the disruption of reproductive functions in both males and females, in addition to the detrimental consequences for the fetus or offspring [27]. According to REACH regulations, the reproductive toxicity evaluation of a substance is fundamentally conducted through three distinct studies: (i) a reproduction/developmental toxicity screening test (such as OECD TG 421/422), (ii) prenatal developmental toxicity studies involving two species, and (iii) extended one-generation reproductive toxicity study (EOGRTS) [28]. The development of alternative *in vitro* models for assessing developmental toxicity in cosmetics is a growing field, driven by ethical concerns and regulatory demands to reduce animal testing. Alternative embryotoxicity-related methods are crucial for reducing reliance on animal testing and improving the predictability of developmental toxicity assessments. Three established alternative embryotoxicity-related methods are currently available for testing: (i) the whole embryo culture (WEC) test, (ii) the micromass (MM) test, and (iii) the embryonic stem cell test(EST). These methods can be employed to identify strong embryotoxic substances (development) and improve mechanistic understanding of developmental toxicity. The development of a novel testing strategy for reproductive toxicity was assigned to the integrated project ReProTect, a consortium established by the European Centre for the Validation of Alternative Methods (ECVAM) [29].

3.5.5.5 Mutagenicity/Genotoxicity

The term "mutagenicity" denotes the induction of permanent transmissible alterations in the quantity or structure of the genetic material of cells or organisms. Such changes may affect a single gene or gene segment, a cluster of genes, or entire chromosomes, with chromosomal effects being either structural (e.g., deletions, translocations) or numerical (e.g., aneuploidy).

Assessment of fragrance ingredient for mutagenicity (genotoxicity) is deemed essential if the substance contains any of the alert structures for this endpoint, or if mutagenicity potential has been suggested by other assessments of the substance or structurally analogous materials [30]. Generally, genotoxic potential is usually investigated initially through *in vitro* assays, in the first instance *in vitro* assays, bacterial reverse mutation assay (Ame's test) for point mutation. Another *in vitro* mammalian cell assay for chromosomal aberration test (e.g., micronucleus assay) to analyze a possible genotoxic risk resulting from a chemical exposure. Nevertheless, Scientific Committee on Cosmetic Products and Non-Food Products (SCCNFP) recommends a three-test battery (Ame's test, *in vitro* mammalian cytogenetic test, and *in vitro* mammalian gene mutation test) to assess the more comprehensive mutagenic/genotoxic properties of a cosmetic ingredient [31]. This integrated approach ensures that potential risks arising from chemical exposure are more accurately identified before widespread consumer use.

The assessment of genotoxicity risk in cosmetics, particularly concerning fragrances, is a multifaceted process that involves various methodologies and regulatory considerations. Fragrances, while providing sensory benefits, can pose health risks such as allergic reactions and potential genotoxic effects. The evaluation of these risks is crucial for consumer safety and involves both traditional and innovative testing methods. However, increasing ethical and regulatory pressures have driven the adoption of new approach methodologies (NAMs), including advanced *in vitro* models and *in silico* predictions, to reduce reliance on animal testing.

3.5.5.6 Carcinogenicity

Carcinogenicity refers to the capacity of a chemical to induce tumor formation, whether benign or malignant, or to enhance their malignancy, or accelerate the onset of tumors [32]. While some carcinogens are well-documented, the mechanisms by which they induce cancer can vary significantly, involving both genotoxic and nongenotoxic pathways. Genotoxic carcinogens directly damage DNA, leading to mutations while nongenotoxic carcinogens do not directly damage DNA but can induce cancer through epigenetic changes or by promoting cell proliferation.

The standard rodent carcinogenicity bioassay (RCB) has been a cornerstone in assessing the carcinogenic potential of chemicals, including those used in cosmetics. This method involves chronically exposing rodents to test substances over a two-year period to observe the potential cancer development [33]. Despite its widespread use, the RCB has faced criticism regarding its relevance and accuracy in predicting human carcinogenicity. The RCB assay is designed to maximize sensitivity, often using maximum tolerated doses (MTD) to ensure that potential carcinogens are not missed. However, this approach can lead to false positives, as high doses may not reflect typical human exposure levels under realistic exposure scenarios [34]. The RCB is resource-intensive, time-consuming, and raises ethical concerns due to the extensive use of animals, and these shortcomings have accelerated the search for NAMs.

With the ban on animal testing for cosmetics in the EU, cell transformation assays (CTAs) have gained prominence as an alternative method for evaluating the safety of cosmetic ingredients. CTAs are used to evaluate the carcinogenic potential of cosmetic ingredients by detecting cellular changes indicative of cancer, such as loss of contact inhibition and anchorage-independent growth [35]. While CTAs offer a promising alternative to animal testing, they do not fully replicate the complexity of *in vivo* carcinogenesis. The assays focus on specific cellular changes and may not capture all aspects of cancer development. Additionally, the interpretation of CTA results requires careful consideration of the assay's limitations and the context of use.

S. no.	*Toxicological endpoint*	*Name*	*Testing method*	*Type*	*OECD guideline*
1.	Skin sensitization	DPRA	Direct peptide reactivity assay	*In chemico*	OECD TG 442C
		KeratinoSens™	KeratinoSens assay	*In vitro*	OECD TG 442D
		hCLAT	Human cell line activation test	*In vitro*	OECD TG 442E
		USens	U-SENS™—Myeloid U937 skin sensitization test	*In vitro*	
		IL8	Interleukin 8 reporter gene assay	*In vitro*	
		LLNA	Local lymph node assay	*In vivo*	OECD TG
2.	Acute toxicity	LD50	Determination of LD50	*In vivo*	OECD TG 402.
		OoC	Organ on chip	*In vitro*	
3.	Skin irritation	Draize Test	Draize test assay	*In vivo*	
		SkinEthic RHE	RHE model	*In vitro*	OECD TG 439
		LabCyte EPI-MODEL24 SIT	LabCyte EPI-MODEL24 skin irritation test	*In vitro*	
4.	Phototoxicity	3T3 NRU	Neutral red uptake assay	*In vitro*	OECD TG 432
		KeraSkin™	KeraSkin™ phototoxicity assay	*In vitro*	OECD TG 498
		RhE	Reconstructed human epidermis (RHE) phototoxicity test method	*In vitro*	OECD TG 498

(*Continued*)

S. no.	*Toxicological endpoint*	*Name*	*Testing method*	*Type*	*OECD guideline*
5.	Reproductive/ Developmental toxicity		Reproduction/ Developmental toxicity screening test		OECD TG 421
			Prenatal developmental toxicity studies		OECD TG 414
		EOGRTS	Extended one generation reproductive toxicity studies		OECD TG 443
		WEC	Whole embryo culture test	*In vitro*	OECD TG
		MM	Micromass test		
		EST	Embryonic stem cell test		
6.	Mutagenicity/ Genotoxicity		Ames test	*In vitro*	
			Micronucleus test	*In vitro*	OECD TG 487
7.	Carcinogenicity	RCB	Rodent carcinogenicity bioassay	*In vivo*	
		Bhas 42 CTA	Cell transformation assay		OECD TG 231

3.6 REGULATORY STATUS OF ALLERGEN LABELING

Based on the viewpoint of the SCCNFP, the predecessor of the Scientific Committee on Consumer Safety (SCCS), a practical administrative ruling in 1999 established specific labeling standards for probable fragrance allergens. These standards included concentration limitations of 0.01% and 0.001% for rinse-off and leave-on products, respectively, ensuring that consumers were adequately informed of potential allergenic risks [36]. Over the time, allergen list (Annex III of Reg. 1223/2009) underwent modifications; notably, two substances were prohibited due to safety concerns: the first was hydroxyisohexyl 3-cyclohexene carboxaldehyde (HICC) in 2017, and the second was buthylphenyl methylpropional in 2021 owing to their carcinogenic, mutagenic, and reprotoxic (CMR) properties [37]. Furthermore, these allergens are specifically included in Annex II of Regulation 1223/2009, which lists substances banned from use in cosmetic products.

A revised assessment of fragrance allergens in cosmetics items was released by the SCCS in 2012 opinion, which went beyond the original 26 listed substances, identifying over 80 individual fragrance substances and 12 natural extracts as potential allergens of concern. This update, was accepted based on emerging toxicological evidence, aimed to strengthen consumer protection by expanding the scope of regulated allergens and improving transparency in cosmetic labeling.

- An extra 56 entries need to be included in Reg. 1223/2009, Annex III.
- Typically, it is recommended to include 27 separate labeling entries for plant essential oils or extracts.
- A few of the current entries in Annex III should be changed (for certain substances, just the peroxide value limitations were assigned; with the revision, their individual labeling should accompany the restrictions).

According to the SCCS opinion, amendment (EU) 2023/1545 adds fragrance allergens needing individual labeling to Annex III of the Cosmetics Regulation. Furthermore, substances such as prehaptens and prohaptens, which can be oxidized or bio-activated and converted to known allergens, should be regarded as equivalent to fragrance allergens and subject to the same labeling requirements (EU 2023/1545).

3.7 RISK MANAGEMENT

- A new self-explanatory symbol for skin sensitizers and irritants should be required for consumer items containing allergens.
- Products containing hazardous substances shouldn't have appealing packaging that makes you think the products are safe.
- Maximum concentrations in the products should be stated based on the usage patterns and sensitivity potentials.
- Although the amendment to fragrance allergen labeling is a positive development, conventional physical labeling methods have practical constraints. One easy approach for users to grasp may be a smart media app on a smartphone. An app like this might scan the label for allergen names and visually provide a rating for how severe, moderate, or mild the allergen is.
- For the purposes of primary prevention, a restriction of fragrance allergens may be sufficient. However, sensitized persons may develop symptoms when they are exposed to lower concentrations of an allergen than the maximum permitted levels. Therefore, as a measure of secondary prevention, it is important to provide information on the presence of individual fragrance allergens in cosmetic products so that sensitized persons can avoid contact with the substance to which they are allergic.

3.8 CONCLUSION

This chapter aims to provide current information on the risk assessment of fragrances. The risk evaluation of fragrance compounds is an effective scientific approach to guarantee the safety of consumer goods. Moreover, risk communication based on risk assessment is a crucial endeavor involving fragrance consumers, regulatory bodies, producers, academics, and the media to foster mutual understanding about product safety. In addition, the coexistence of regulatory needs and recent developments in toxicological testing protocols for fragrance materials allows for proper hazard and potency estimation and risk management. This chapter serves as a catalyst for the current review of the risk assessment methodologies with toxicological testing along with allergen labelling legislations and its management which strengthened the usefulness for skin sensitization potential and potency estimation and further prevention of allergic diseases.

REFERENCES

1. Basketter, D.A., J. Huggard, and I. Kimber, *Fragrance inhalation and adverse health effects: The question of causation.* Regul Toxicol Pharmacol, 2019. **104**: pp. 151–156.
2. Arnesdotter, E. et al, *An overview of current practices for regulatory risk assessment with lessons learnt from cosmetics in the European Union.* Crit Rev Toxicol, 2021. **51**(5): pp. 395–417.
3. Lee, I. et al, *Assessment of the skin sensitization potential of fragrance ingredients using the U-SENS™ assay.* Toxicol In Vitro, 2022. **79**: p. 105298.
4. van Amerongen, C.C.A. et al, *Skin exposure to scented products used in daily life and fragrance contact allergy in the European general population - the EDEN fragrance study.* Contact Dermat, 2021. **84**(6): pp. 385–394.
5. Tramontana, M. et al, *Advancing the understanding of allergic contact dermatitis: From pathophysiology to novel therapeutic approaches..* Front Med (Lausanne), 2023. **10**: p. 1184289.
6. Nishijo, T. et al, *Application of the dermal sensitization threshold concept to chemicals classified as high potency category for skin sensitization assessment of ingredients for consumer products.* Regul Toxicol Pharmacol, 2020. **117**: p. 104732.
7. COMMISSION, E.; Available from: https://health.ec.europa.eu/system/files/2016-11/cosmetic_1223_2009_regulation_en_0.pdf.
8. Work, E.A.S.H.; Available from: https://osha.europa.eu/en/legislation/directives/regulation-ec-no-1907-2006-of-the-european-parliament-and-of-the-council.
9. Kolle, S.N., R. Landsiedel, and A. Natsch, *Replacing the refinement for skin sensitization testing: Considerations to the implementation of adverse outcome pathway (AOP)-based defined approaches (DA) in OECD guidelines.* Regul Toxicol Pharmacol, 2020. **115**: p. 104713.
10. Aleksic, M. et al, *The skin sensitization adverse outcome pathway: Exploring the role of mechanistic understanding for higher tier risk assessment.* Crit Rev Toxicol, 2024. **54**(2): pp. 69–91.
11. OECD, *Test No. 442C: In Chemico Skin Sensitisation.* 2024.
12. Ferreira, I. et al, *Development of an in chemico high-throughput screening method for the identification of skin sensitization potential.* Arch Toxicol, 2023. **97**(9): pp. 2441–2451.
13. OECD, *Test No. 442D: In Vitro Skin Sensitisation.* 2024.

14. Settivari, R.S. et al, *Application of the KeratinoSens™ assay for assessing the skin sensitization potential of agrochemical active ingredients and formulations.* Regul Toxicol Pharmacol, 2015. **72**(2): pp. 350–60.
15. OECD, *Test No. 442E: In Vitro Skin Sensitisation.* 2024.
16. Tanabe, I. et al, *Development of an in vitro sensitisation test using a coculture system of human bronchial epithelium and immune cells.* Altern Lab Anim, 2023. **51**(6): pp. 387–400.
17. Na, M. et al, *Weight of evidence approach for skin sensitization potency categorization of fragrance ingredients.* Dermatitis, 2022. **33**(2): pp. 161–175.
18. Thélu, A., S. Catoire, and S. Kerdine-Römer, *Immune-competent in vitro co-culture models as an approach for skin sensitisation assessment.* Toxicol In Vitro, 2020. **62**: p. 104691.
19. Gądarowska, D. et al, *Alternative methods for skin-sensitization assessment.* Toxics, 2022. **10**(12): p. 740.
20. Gerberick, G.F. et al, *Local lymph node assay (LLNA) for detection of sensitization capacity of chemicals.* Methods, 2007. **41**(1): pp. 54–60.
21. OECD, *Test No. 429: Skin Sensitisation.* 2010.
22. Xavier, J., et al., *23 - organ-on-a-chip for toxicity evaluation*, in *Biomedical product and materials evaluation*, P.V. Mohanan, Editor. 2022, Woodhead Publishing. pp. 611–633.
23. Klein-Szanto, A.J.P., and C.J. Conti, *29 - skin and oral mucosa*, in *Handbook of toxicologic pathology* (Second Edition), W.M. Haschek, C.G. Rousseaux, and M.A. Wallig, Editors. 2002, Academic Press. pp. 85–119.
24. OECD, *Test No. 439: In Vitro Skin Irritation: Reconstructed Human Epidermis Test Method.* 2021.
25. Kim, K., H. Park, and K.M. Lim, *Phototoxicity: Its mechanism and animal alternative test methods.* Toxicol Res, 2015. **31**(2): pp. 97–104.
26. Kandarova, H., and Liebsch M., *Chapter 35 - The EpiDerm™ phototoxicity test (EpiDerm™ H3D-PT)*, in *Alternatives for dermal toxicity testing*, C. Eskes, E. van Vliet, and H.I. Maibach, Editors. 2017, Springer International Publishing. pp. 483–503.
27. Nikolaidis, E., *Chapter 6 - Relevance of animal testing and sensitivity of end points in reproductive and developmental toxicity*, in *Reproductive and developmental toxicology* (Third Edition), R.C. Gupta, Editor. 2022, Academic Press. pp. 93–106.
28. Pistollato, F. et al, *Current EU regulatory requirements for the assessment of chemicals and cosmetic products: Challenges and opportunities for introducing new approach methodologies.* Arch Toxicol, 2021. **95**(6): pp. 1867–1897.
29. Silva, R.J., and S. Tamburic, *A state-of-the-art review on the alternatives to animal testing for the safety assessment of cosmetics.* Cosmetics, 2022. 9.
30. Ford, R.A. et al, *Criteria for development of a database for safety evaluation of fragrance ingredients.* Regul Toxicol Pharmacol, 2000. **31**(2 Pt 1): pp. 166–81.
31. Di Sotto, A. et al, *Genotoxicity assessment of some cosmetic and food additives.* Regul Toxicol Pharmacol, 2014. **68**(1): pp. 16–22.
32. Madannejad, R. et al, *Toxicity of carbon-based nanomaterials: Reviewing recent reports in medical and biological systems.* Chem Biol Interact, 2019. **307**: pp. 206–222.
33. Suarez-Torres, J.D., F.A. Jimenez-Orozco, and C.E. Ciangherotti, *The 2-year rodent bioassay in drug and chemical carcinogenesis testing: Sensitivity, according to the framework of carcinogenic action.* Toxicol Mech Methods, 2020. **30**(6): pp. 462–475.

34. Suarez-Torres, J.D., C.A. Orozco, and C.E. Ciangherotti, *The 2-year rodent bioassay in drug and chemical carcinogenicity testing: Performance, utility, and configuration for cancer hazard identification*. J Pharmacol Toxicol Methods, 2021. **110**: p. 107070.
35. Colacci, A. et al, *The cell transformation assay: A historical assessment of current knowledge of applications in an integrated approach to testing and assessment for non-genotoxic carcinogens*. Int J Mol Sci, 2023. **24**(6): p. 5659.
36. Bialas, I., S. Zelent-Kraciuk, and K. Jurowski, *The skin sensitisation of cosmetic Ingredients: Review of actual regulatory status*. Toxics, 2023. **11**(4): p. 392.
37. Johansen, J.D., *Fragrance contact allergy: A clinical review*. Am J Clin Dermatol, 2003. **4**: pp. 789–798.

Chapter 4

Biodegradability and eco-toxicological impacts of cosmetic fragrances

Abhay Bajaj
Environmental Toxicology Group, FEST Division, CSIR-Indian Institute of Toxicology Research, Lucknow, India

4.1 INTRODUCTION

Cosmetics, under the Drugs and Cosmetics Act 1940 (https://cdsco.gov.in/opencms/opencms/en/Acts-and-rules/Drugs-and-Cosmetics-Act/), are defined as "*any article intended to be rubbed, poured, sprinkled or sprayed on, or introduced into, or otherwise applied to, the human body or any part thereof for cleansing, beautifying, promoting attractiveness or altering the appearance.*" Fragrance compounds impart pleasing scents to household and cosmetic products, enhancing the consumer's experience. Fragrances not only solve functional problems, such as when used as additives or preservatives, but they can also enhance well-being by providing freshness and alleviating stress. More than hundred of fragrances are generated globally every year, and consumer research shows fragrance as one of the key factors in deciding preference for cosmetic and personal care products.

Fragrance compounds can be classified as natural and synthetic (Table 4.1). Due to their source of origin, these fragrances vary considerably in their biodegradation (Pastor-Nieto & Gatica-Ortega, 2021). Natural fragrance compounds like floral extracts, resins, and essential oils are derived from plants. They often break down relatively quickly in the environment and are less toxic, making them eco-friendly. Synthetic fragrance compounds are derived from chemicals or petrochemicals like ethylhexyl methoxycinnamate, phthalates, and synthetic musks. They are complex molecules with high persistence in the environment for extended periods, often leading to ecotoxicity. Generally, natural fragrances are comparatively more biodegradable than their synthetic counterparts; depending on specific chemical composition, some may still contain compounds that degrade slowly or persist in the environment. For instance, plant-derived citrus oils are readily biodegradable, while plant-derived sandalwood degrades gradually due to sesquiterpenes. Therefore, the source of origin and chemical composition are equally crucial for generating eco-friendly fragrances and cosmetic products. Synthetic fragrances are commonly used in cosmetics due to their stability, cost-effectiveness, and ability to create various scents. However, many of these fragrance compounds are poorly biodegradable, persist in the environment, and may

DOI: 10.1201/9781003558477-4

Table 4.1 Major classes of fragrance compounds and their eco-toxicological effects

Class	*Compounds*	*Feature*	*Environmental persistence/ bioaccumulation*	*Eco-toxicological risks*
Polycyclic	Galaxolide, tonalide	Synthetic	High	Very toxic to aquatic life (acute and chronic)
Nitro	Musk xylene, musk ketone			
Macrocyclic and alicyclic	Civetone, ambrettolide			
Terpenes	Limonene, linalool	Natural components of essential oils	Low	Exhibit toxic effects at high concentration
Aldehydes and Ketones	Cinnamaldehyde, citral, vanillin	Distinct aromatic properties		Exhibit toxic effects on aquatic organisms
Phenolic Compounds	Eugenol, vanillin	Derived from natural sources	Moderate	Exhibit eco-toxicological effects due to their aromatic nature
Esters	Benzyl acetate, linalyl acetate	Fruity and floral fragrances	Moderate	Pose risks to aquatic ecosystems

even bioaccumulate, especially in aquatic ecosystems. Further, some natural and synthetic fragrance compounds can be toxic to aquatic species in higher concentrations and may pose potential long-term ecological risks. In many cosmetic products, fixatives are also added to stabilize the scent over some time. However, these fragrance fixatives exhibit slower biodegradation and, therefore, have higher persistence, contributing to long-term environmental damage. The chapter highlights commonly used fragrances in cosmetics, their biodegradability and eco-toxicological impact on different ecosystems, and guidelines or frameworks implemented globally to regulate biodegradability and eco-toxicological risks of fragrance compounds. Due to rising health concerns and environmental awareness, the cosmetic industry has a rising demand to develop and use fragrance compounds that are biodegradable and less toxic to all species, hence eco-friendly.

4.2 BIODEGRADATION OF FRAGRANCES: FACTORS AND DRIVERS

Fragrance compounds are extensively used consumer products that mostly end up in wastewater, often sorbed to the wastewater/sewage sludge. This way, these compounds enter the environment, and their persistence and bioaccumulation determine their long-term effect on the environment. Under environmental conditions, abiotic processes like photolysis, atmospheric oxidation, hydrolysis, and biotic processes like biological degradation account for the degradation of chemicals, thereby affecting the persistence of chemicals. Generally, in the natural environment, microbes, including bacteria and fungi, degrade chemicals into simpler substances like water and carbon dioxide, and this process is called biodegradation. Biodegradation is vital in minimizing the long-term environmental impact of harmful chemicals, including synthetic fragrance compounds. Depending on the persistence of chemicals and the time taken to break down the chemicals, biodegradability can be classified as *readily biodegradable*—a substance that breaks down quickly and entirely within a short period, i.e., 60% degradation within 28 days under stringent test conditions, *inherently biodegradable*—a substance that takes longer time but eventually breaks down under favorable conditions, and *non-biodegradable*—a substance that does not break down quickly and may persist in the environment for extended periods. The environmental fate of fragrance compounds can further be assessed and documented as persistent or bio-accumulative based on biodegradation under set guidelines and standards discussed later in this chapter. Generally, physicochemical properties like hydrophobicity, volatility, and inhibition potential of fragrance compounds at high test concentrations affect the biodegradation of fragrances during screening tests performed under standard guidelines. Reducing stringency for specific guidelines, such as extended incubation, can also affect the biodegradability of the test fragrance compound. One study reviewing quaternary carbon-containing compounds under high (OECD 301F) and low (OECD 301D) concentrations and extended incubations showed that ionones were generally biodegradable within 28 days or readily biodegradable. In contrast, damascones only reached ultimate biodegradation in OECD 301D tests and required extended incubations (Dick et al., 2016). Overall, chemical complexities owing to molecular size, chemical structure, functional groups, and environmental conditions can profoundly influence the biodegradability of fragrance compounds. Generally, fragrance compounds with smaller and simpler structures undergo microbial degradation, unlike complex large molecules, which may persist, eventually accumulating in soil or aquatic systems like synthetic polycyclic musks. In an aqueous environment, the accessibility of certain hydrophobic fragrance compounds is often limited to microbes, and they may even persist in aquatic systems or bioaccumulate in aquatic organisms. Further, certain volatile fragrance compounds are not biodegradable in water environments and may evaporate into the atmosphere. These volatile fragrance compounds and non-readily biodegradable fragrance ingredients, such as pamplewood, may degrade through other environmental processes, such as photodegradation in sunlight. Indirect photodegradation studies indicated this compound is intrinsically photolabile (Lin & Emberger, 2017). At the same time, other studies showed direct photodegradation of certain fragrance compounds like monocyclic musk tibetene (Canterino et al., 2008).

Fragrances with complex aromatic rings, alkyl chains, or multiple functional groups often resist biodegradation. The presence of different functional groups also influences biodegradation. For instance, in natural and synthetic fragrances, ester is a common functional group that is often easier to break down by microbes. However, aromatic rings in commonly used fragrance ingredients like alkyl phenols, musk ketones, or musk xylene are more resistant due to their hydrophobic nature and persist in water bodies, posing a risk of bioaccumulation and toxicity, thus disrupting aquatic ecosystems. At the same time, phthalates like diethyl phthalate and dibutyl phthalate, commonly used as fixative in synthetic fragrances to increase their longevity, are difficult to degrade and often accumulate in the environment.

Typically, synthetic fragrance compounds have poor biodegradability due to their chemical complexity, leading to long-term environmental impact and pollution. In contrast, natural fragrances with smaller and simpler chemical compositions, like terpenes or terpenoids, are generally more biodegradable; however, those with complex molecular structures like sesquiterpenes may degrade more slowly. The concentration of fragrance compounds can also affect biodegradation; at lower concentrations, it may biodegrade quickly, while at higher concentrations, the natural breakdown process may be hindered, causing longer persistence in the environment. Besides, environmental factors like temperature, pH, oxygen, and moisture influence the biodegradability of these compounds; however, under ecological conditions, the microbial diversity and adaptation of specific microbes also influence biodegradation to a greater extent. Certain microbes evolve and adapt efficiently to break down environmental chemicals, thus enhancing biodegradation. However, certain newer synthetic chemicals and emerging pollutants can persist in the environment as microbes may not yet be adapted to have the capability to degrade them. Additionally, the coexistence of certain environmental pollutants like heavy metals, industrial chemicals, and other chemical contaminants can decrease biodegradation efficiency by detrimentally affecting microbial communities and stressing or inhibiting microbial processes.

4.3 GUIDELINES FOR TESTING THE BIODEGRADABILITY OF FRAGRANCES

Fragrance compounds in cosmetics significantly affect the environment. Therefore, testing and standards for biodegradability of such compounds is essential for environmental risk assessment and regulation (Dick et al., 2016). These standard tests assess how quickly and efficiently fragrance compounds break down into non-toxic components under specific environmental conditions and do not persist in the environment. Biodegradability testing also adheres to regulatory compliance norms of respective regions. These further enable cosmetic products to qualify for eco-label certifications (Table 4.2) that require biodegradable and non-toxic ingredients. Modern analytical

Table 4.2 Certifications and eco-labels signifying environmental friendliness of a product

Label type	*Significance*
EU Ecolabel	Voluntary label for environmental excellence. Tackles the main environmental impacts of products along their full lifecycle, from extraction of raw material to disposal. Requires all chemical components, including fragrances, to meet strict biodegradability standards. Synthetic musks and other persistent compounds are often excluded.
COSMOS Organic	Promotes the use of natural or naturally derived ingredients with proven biodegradability. Products with this certification avoid harmful synthetic chemicals. Promotes green chemistry and organic agriculture.
Cradle-to-Cradle Certification	Ensures sustainability performance regarding safety, circularity, and responsibility for materials and products. Ensures ingredients are biodegradable or recyclable, minimizing their environmental footprint.

Note: Biodegradability testing is integral to achieving eco-certifications.

techniques also provide precise data on the biodegradability of fragrance compounds (Kempińska-Kupczyk & Kot-Wasik, 2019). For instance, high-performance liquid chromatography (HPLC) gives real-time tracking of compound breakdown; mass spectrometry helps in the identification of intermediate toxic by-products of biodegradation; stable isotope analysis assesses the complete mineralization of a substance, and biodegradation modeling using computational tools predicts biodegradability based on the molecular structure (Singh et al., 2021).

Further, key testing standards involve *ready biodegradability* tests to evaluate the degradation of a compound under optimal aerobic conditions within a short period. These laboratory tests assess the readiness of a compound to biodegrade in natural environments. The Organisation for Economic Co-operation and Development (OECD) provides a framework and guidelines for biodegradability testing of chemicals (Dick et al., 2016) (https://www.oecd.org/en/topics/sub-issues/testing-of-chemicals/test-guidelines.html).

According to the *OECD 301* test (OECD 1992; Api et al., 2021a), a compound is readily biodegradable if 60–70% of the substance is degraded within 28 days under aerobic conditions. There are certain variations in the methodology based on indicators of biodegradation. For instance, the carbon dioxide evolution test (Test 301B) measures the CO_2 generated during microbial breakdown. The modified MITI test (Test 301C) measures oxygen consumption, while the manometric respiratory test (Test 301F) measures oxygen depletion in a closed system as microbes degrade/consume the test substance. OECD 301 tests generally assess simple and water-soluble compounds like esters and alcohols. Complex compounds like synthetic musks often fail due to their persistence property. Inherent biodegradability test standards are followed for such chemical compounds that fail ready biodegradability tests and may degrade over a more extended period or under less optimal conditions. According to the *OECD 302 test* (OECD 2009; Api et al., 2022), extended testing over weeks or months is performed to determine the biodegradation of a substance. For example, polycyclic musks (PCMs) with shorter persistence periods may degrade to some extent under prolonged testing. Certain compounds undergo degradation under oxygen-free conditions in various environments like landfills, sediments, or deep water; anaerobic biodegradability tests are performed for such compounds. Under test guideline *OECD 310* (OECD 2014), a compound is tested for biodegradation potential under oxygen-limited conditions. These tests are essential for synthetic fragrances that degrade slowly under anaerobic conditions, causing risk in wastewater or sewage sludge (Api et al., 2021b).

Although these tests evaluate the biodegradation of test substances, there are specific challenges (Dick et al., 2016). Cosmetics or fragrance blends usually contain complex compounds with varying biodegradability, making testing of selected compound difficult. Further, laboratory tests simulate optimal conditions but may not reflect actual environmental conditions where additional intermediates may be formed. Thus, additional testing may be required for some compounds to assess the environmental impact and toxicity of such intermediates. Testing standards for biodegradability are crucial for regulating and minimizing the environmental impact of fragrance compounds. Thus, advancements in testing methodologies are required to adopt safer and more biodegradable compounds in the fragrance industry.

4.4 MICROBIAL DEGRADATION OF FRAGRANCE COMPOUNDS

The biodegradation of fragrance molecules by microorganisms is governed by the distribution and bioavailability of the molecules to microorganisms, as well as their intracellular transport (Zhang et al., 2013a). Another factor that can affect and potentially hinder the degradation is the structural properties of synthetic musks, which are often persistent, and certain specific activities, such as antimicrobial activity, as exhibited by some of the fragrance molecules like linalool. Unlike several fragrance compounds, terpenes like limonene and linalool are readily biodegradable. Rhodococci *R. erythropolis* DCL14 and *R. opacus* PWD4 were demonstrated to transform limonene to trans-carveol by an NADPH-dependent hydroxylation reaction (van der Werf et al., 1999; Duetz et al., 2001). Biotransformation of limonene into α-terpineol, a commercially valuable monoterpene, by fungi *Penicillium digitatum* and *Fusarium oxysporum* as well as by bacteria *Pseudomonas gladioli*, *P. fluorescens*, and *Sphingobium* sp. has also been reported (Cadwallader et al., 1989; Bicas et al., 2009; Molina et al., 2019).

Of all the global fragrance compounds, synthetic musk fragrances, specifically PCMs, represent 95% of the compounds, hinting at their large production and utilization volumes. Synthetic musks have garnered attention as emerging contaminants due to their persistence and inadequate removal by current treatments in wastewater treatment plants (WWTPs) (Arruda et al., 2022). Therefore, most synthetic musks end up in wastewater, and during biological wastewater treatment, adsorption to the sludge is the major removal pathway, while biotransformation/biodegradation is negligible. Even during anaerobic sludge stabilization, no removal was observed, suggesting that treated sludge is the major sink for PCMs (Clara et al., 2011). Little data is available on studies focusing on degradation of fragrance compounds by individual microorganisms. Most studies have reported using activated sludge to study biodegradation, often without any details on the characterized taxonomical composition of the microbial inoculum. Dick et al. (2016) used activated sludge from a sewage treatment plant to evaluate the biodegradability of 48 fragrance compounds with a primary focus on assessing the limitations of biodegradation tests. Similarly, nitro-musks, known for their various eco-toxicological effects, have been found in various environmental matrices (Gao et al., 2019). Nitro-musks like musk xylene were shown to be non-biodegradable in aerobic environments (Tas et al., 1997). However, their nitro groups can undergo reduction reactions under anaerobic conditions, creating 2/4-amino-musk xylenes (Käfferlein et al., 1998). Similarly, biodegradability tests performed under OECD Test Guideline (TG) showed that musk xylene (TG 301C) (ECB, 2005a),, musk ketone (TG 302C) (ECB, 2005b) and musk moskene (TG 302C) (OSPAR Commission, 2004) showed 0% degradation in 28 days and were not readily biodegradable. However, studies also reported that musk xylene and musk ketones undergo primary degradation in sewage treatment plants, water, and soil; with degradation involving reduction of a nitro group to an amine group (ECB, 2005a; 2005b). Alarmingly, the amine degradation products appear persistent in the environment (Rimkus et al., 1999; Herren and Berset, 2000; Osemwengie and Steinberg, 2001) and are expected to have moderate to high lipophilicity and toxicity (OSPAR Commission, 2000; US EPA, 2008; 2012). Such synthetic fragrance compounds are alarmingly prevalent in the environment and are widely reported. Tasselli et al. (2023) investigated the presence of PCMs in water and sediment samples of tributaries of a lake in Northern Italy, revealing the distribution and fate of PCMs in freshwater ecosystems. Another recent study reported the prevalence of selected six synthetic polycyclics and five nitro musk compounds across 43 WWTPs in the Czech Republic. The study also assessed the effect of WWTP technology parameters on synthetic musk compound content in sewage sludge. It was found that in all samples tested, galaxolide (HHCB) and tonalide (AHTN) were predominant synthetic polycyclic musk compounds (SPMCs) and it did not change over a gap of two years in the same samples. Interestingly, the study highlighted their persistence in the environment and the limitation of current common WWTP technologies to affect the accumulation of musk compounds in sewage sludge

(Košnář et al., 2021). Generally, PCMs like galaxolide and tonalide are considered resistant to biodegradation. However, their biotransformation by fungi in freshwater systems was also explored. Martin et al. (2007) reported the production of hydroxylated derivatives and other metabolites such as diketone, peroxide, and O-methylated derivatives primarily through extracellular oxidoreductases such as laccase activity by freshwater fungi *Myrioconium* sp. strain UHH 1-13-18-4 and *Clavariopsis aquatica* strain WD(A)-00-1. The study reported that laccases isolated from selected fungi could oxidize HHCB and AHTN and catalyze the formation of the lactone-type metabolite. Interestingly, the study highlights that for addressing the fate of PCMs in freshwater, the eco-toxicological effect of metabolites formed due to fungal biotransformation in freshwater environments may be considered (Martín et al., 2007). Similarly, the removal of SPMC from activated sludge in wastewater treatment systems has been explored. In a wastewater treatment plant in Milan, Italy, study showed 20% and 50% removal rates for HHCB and AHTN respectively, from wastewater primarily due to adsorption on activated sludge rather than biodegradation (Tasselli et al., 2021). In another study, microbe-assisted phytoremediation of HHCB and AHTN was reported. This study focused on the bioremediation of HHCB and AHTN in biosolid-amended soil by white rot (lignolytic) fungi, namely *P. ostreatus, I. lacteus*, and *P. chrysosporium*, combined with phytoremediation and without phytoremediation. The study reported that phyto-mycoremediation treatment showed improved removal of galaxolide by 46.3% and tonalide by 34.5%, which was only 9% under the natural attenuation process (Chane et al., 2023).

Interestingly, studies report the persistence of synthetic fragrance compounds in the environment; however, biotransformation can also be found by using selected microbes under natural conditions (Table 4.3). When used in cosmetics or similar formulations, the biodegradability of fragrance compounds under natural conditions depends on multiple interacting factors, as discussed in Section 4.2. These include the chemical structure of the fragrance compound, its concentration, the environmental conditions, the microbial communities, and the presence of additives or fixatives. Taking cues from the persistence of synthetic musks and related compounds, structure-activity relationships can be exploited to develop environmentally friendly fragrances that break down efficiently and do not persist, leading to potential toxicity in the environment. Employing computational predictive tools like ecological structure activity relationships (ECOSAR) (US EPA 2012), the aquatic toxicity of these compounds can be estimated. Similarly, quantitative structure biodegradability relationship (QSBR) models for predicting biodegradation rate of organic chemicals (Acharya et al., 2019) can be developed for fragrance molecules which is an essential parameter for environmental risk assessment. Intriguingly, computational tools and microbial ecology advances can provide sustainable solutions to manage fragrance-related environmental impacts.

Table 4.3 Specific microbes for biodegradation of fragrance compounds

Microbes	*Degradation*	*Reference*
1. Microbial degradation of terpenes (e.g., limonene, linalool)		
Terpenes, derived from plants, are biodegradable under aerobic conditions.		
Pseudomonas sp.	Commonly found in soil and water, efficient at degrading limonene and linalool.	Chatterjee & Bhattacharyya (2001)
Rhodococcus sp.	Known for its ability to degrade hydrophobic and volatile compounds like limonene. Converts limonene to carveol, carvone. Produces enzymes like limonene monooxygenase to initiate the breakdown.	Duetz et al (2001)
2. Degradation of synthetic musks		
Synthetic musks, especially nitro and polycyclic musks, are resistant to biodegradation due to their complex and stable structures. However, oxidation of aromatic rings followed by cleavage of the structure can be a plausible mechanism for microbial degradation of aromatic compounds.		
Rhodococcus sp. *Bacillus* sp.	Found in wastewater and soil. Effective against polycyclic musks like galaxolide and tonalide.	Azaroff et al. (2021)
3. Degradation of aldehydes and ketones		
Aldehydes and ketones are simpler molecules used in fragrances (e.g., cinnamaldehyde, vanillin). These compounds are more biodegradable compared to synthetic musks.		
Acinetobacter sp.	Degrades benzaldehyde via dehydrogenase activity. Commonly found in wastewater and activated sludge.	MacKintosh & Fewson (1988)
Cystobasidium sp.	Efficiently degrade vanillin to vanillyl alcohol.	Rönnander et al. (2018)
Pseudomonas sp.	Efficiently degrade vanillin to protocatechuate.	Priefert et al. (1997)
4. Degradation of phenolic fragrance compounds		
Phenolic compounds (e.g., eugenol, vanillin) are commonly used in fragrances and degrade through oxidation pathways.		
Pseudomonas sp.	Utilizes eugenol hydroxylase enzymes for initial oxidation and degrades eugenol to coniferyl alcohol and further metabolizes it to vanillic acid.	Priefert et al. (1999)

Note: Microbial degradation of fragrance compounds involves specific microorganisms capable of breaking down these chemicals into simpler, non-toxic components via enzymatic processes. Chemical structure of the fragrance (e.g., terpenes, musks, aldehydes) and environmental conditions (aerobic or anaerobic) are major attributing factors which affect the degradability of these compounds.

4.5 ECO-TOXICOLOGICAL IMPACT OF FRAGRANCE COMPOUNDS

Fragrance compounds in cosmetics impart various eco-toxicological effects upon release into the environment due to their presence in wastewater or other water sources (Klaschka & Kolossa-Gehring, 2007). For instance, musk fragrances are emerging contaminants detected in the environment and are commonly found in various consumer products (Arruda et al., 2022). The impact of these fragrances widely used in personal care products or household items, especially on aquatic systems, is of high concern due to their persistence in water sources and underexplored effects on microbial communities (Arruda et al. 2022). Biofilms commonly formed in aquatic systems play a significant role in microbial exposure to fragrance compounds. There are three major factors: persistence, bioaccumulation, and toxicity (PBT), influencing the eco-toxicological impacts of fragrances (Zhang et al., 2013a). Persistence can be termed as the occurrence of compounds in the environment for more extended periods due to their resistance to biodegradation, like the majority of synthetic fragrance compounds often persist in the environment, while bioaccumulation of fragrance compounds can be defined as accumulation in the aquatic organisms thereby moving up the food chain; and toxicity refers to disruption of biological processes both in aquatic and terrestrial organisms.

4.5.1 Aquatic toxicity: Persistence and bioaccumulation

Many of the fragrance compounds in cosmetics, both synthetic and natural, can be toxic to aquatic ecosystems due to their lethal effects, persistence, and bioaccumulation in the aquatic environment. For instance, musk compounds (Zhang et al., 2013b; Arruda et al., 2022) and polycyclic musks like galaxolide and tonalite are of great concern due to their resistance to biodegradation and potential for bioaccumulation and, thus, are highly persistent in the aquatic environment. Further, studies have detected their presence in the tissues of fishes, marine mammals, and human breast milk, indicating their accumulation in the food chain. These compounds accumulate in the fatty tissues of aquatic species, and bioaccumulation is combined with biomagnification, increasing their concentrations as they move up to higher levels in the food chain, eventually leading to ecotoxicity (Zhang et al., 2013b). Exposure and accumulation of these musk compounds also lead to lethal effects on growth and reproduction in aquatic species like fish (*Pimephales promelas*), algae (*Pseudokirchneriella subcapitata*), and freshwater mussels (*Lampsilis cardium*) (Gooding et al., 2006). It sometimes leads to sublethal effects like hormonal disruption and behavioral changes. Hormonal disruption or interference with the hormonal systems further affects reproduction and development in aquatic organisms. Terpenes, natural ingredients of essential oils, are more biodegradable than synthetic musks but still possess ecotoxic effects at high concentrations. Limonene (a citrus-derived scent) can cause skin irritation or allergic reactions in aquatic species at elevated levels. Limonene has high acute toxicity as reflected by an LC50 (lethal concentration for 50% of the population) of 0.70 mg/L in *Daphnia magna* (water flea) (Aalizadeh et al., 2017; Ravichandran et al., 2018). Aldehydes and ketones are also widely used in cosmetics due to their distinct aromatic properties, but simultaneously, can be potentially toxic to aquatic organisms. Cinnamaldehyde and citral have been reported for acute toxicity to aquatic organisms such as algae and fish. A study reported that citral has an EC50 (effective concentration causing a 50% reduction in population growth) of 3.1 mg/L in *Pseudokirchneriella subcapitata* (Lotfi et al., 2022). Further, it can cause sublethal effects like disrupted enzyme systems in fish and invertebrates and oxidative stress due to reactive aldehyde groups. Phenolic compounds like eugenol and vanillin are derived from natural sources but can potentially cause eco-toxicological effects because of their aromatic nature and moderate environmental persistence. Eugenol and vanillin are toxic to algal species, disrupting photosynthesis and growth (Gueretz et al., 2017). Further, behavioral changes are also reported, like exposure to vanillin, which have been linked to changes in predator-prey dynamics in aquatic systems. Esters like benzyl acetate and linalyl acetate are commonly used for their fruity and floral fragrances. However, it is moderately biodegradable but can still pose risks to aquatic life. Benzyl acetate has been reported as toxic to aquatic species, and its high concentrations can disrupt microbial communities in wastewater treatment plants (Api et al., 2015). Phenoxyethanol, commonly used as a preservative in many cosmetic products, has been reported to be toxic to aquatic species, reducing water quality and ecosystem health (Duan et al., 2018). In aquatic organisms, exposure to phthalates is also linked to altered sex ratios and reproductive health issues (Zhang et al., 2021b).

Some fragrance compounds, especially synthetic musks and pesticide-like ingredients, are highly persistent in marine and soil environments, often increasing their likelihood of accumulating in aquatic sediments and thus affecting long-term ecosystem functions. Further, these fragrance compounds can migrate through wastewater treatment systems, resulting in high concentrations of fragrance chemicals in aquatic environments, exacerbating their toxic effects (Li et al., 2018). The continuous release of persistent fragrance compounds can disrupt the balance of microbial communities in soil, sediments, or wastewater treatment plants, thereby affecting microbial diversity and impairing functions like nutrient cycling and decomposition. These fragrance compounds disrupt different biochemical and structural processes in microbial populations, such as cell membranes and cytoplasm modifications, leading to cell conformation changes and cell death due to compromised microbial metabolism. For instance, tea tree oil affects the fatty acids profile of the cell membrane in *Escherichia coli*, causing bacteria to die without lysing the cell but altering its structure. Lipid biosynthesis is the key target for essential oils, and bacterial quorum sensing is also disrupted due to the antimicrobial activity of fragrance compounds (Nazzaro et al., 2013; Musilova et al., 2016).

The fragrance compounds like synthetic musks and aldehydes affect specific aquatic species like fishes and disrupt endocrine systems, affecting reproduction and development (Ashcroft et al., 2024). In invertebrates, *Daphnia magna* and other crustaceans are particularly sensitive to synthetic musks and terpenes. Further, acute toxicity leads to mortality or sublethal effects, such as altered swimming behavior. In algae and phytoplankton, such fragrance compounds can restrain chlorophyll and inhibit photosynthesis (Broccoli et al., 2021; Jing et al., 2022). Also, reduced algal growth can affect the entire aquatic food webs.

4.5.2 Terrestrial toxicity: Accumulation and endocrine disruption

During development, certain fragrance compounds interact with endocrine-disrupting chemicals, reprogramming hormonal receptors, leading to heightened sensitivity in later stages of life. Exposure to such environmental toxins during early development alters hormonal signals, causing biological disruption, and in some cases, exposure to environmental epigenetic toxins leaves long-lasting effects on progeny (Wilkinson et al., 2016). Certain synthetic musks and phthalates found in fragrances exhibit endocrine disruption, which can interfere with the hormonal systems of both humans and wildlife (Ashcroft et al., 2024). Personal care products and perfumes may also contain certain endocrine-disrupting chemicals like parabens, and phthalates, which can have a synergistic effect and generate substantial disruptive impacts. This endocrine disruption or hormonal imbalance can lead to various organisms' developmental, reproductive, and metabolic disorders. Fragrance compounds such as anethole from fennel, linalool from lavender, terpinen-4-ol from tea tree oil, and geraniol/nerol from citronella and other flowers have shown signs of endocrine disruption but require very high doses for prolonged period to cause obeservable effects. Essential oils themselves have a low binding affinity for hormone receptors. However, essential oil based personal care products and perfumes containing essential oils may also have synthetic endocrine-disrupting compounds like parabens, phthalates, and xenoestrogens from soy products, which can lead to more substantial disruptive effects. Further, studies have reported that these compounds, when released into the environment, can accumulate in the food chain, thus impacting species at multiple trophic levels. The fragrance compounds in cosmetics also affect soil health, mainly when treated wastewater is applied to agricultural land (Zhang et al., 2013a). Fragrance compounds that are not fully biodegraded and are persistent in the environment can accumulate in the soil, potentially affecting soil organisms and plants that depend on healthy soil ecosystems. In soil microbial communities, fragrance compounds can inhibit microbial respiration and enzymatic activity, affecting nutrient cycling. For instance, terpenes can affect soil microbial communities by inhibiting growth or enzymatic activity (Adamczyk et al., 2013). Further, phenolic and terpene-based fragrances may persist in soil, leading to long-term disruption of microbial communities. Also, high concentrations of volatile terpenes and aldehydes can deter pollinators or interfere with their olfactory navigation. Many fragrance compounds are reported to induce oxidative stress in cells, which can trigger signaling pathways in different cell types (Jin et al., 2010). Oxidative stress further leads to DNA damage, enzyme inhibition, and reduced cellular function. Phthalate esters commonly used in perfumes are reported for their genotoxic effects, causing DNA damage, thus suggesting that their use as solvents or fixatives should be regulated (Al-Saleh et al., 2017).

4.6 TESTING THE ECO-TOXICOLOGICAL IMPACT OF FRAGRANCE COMPOUNDS

Due to environmental and health concerns caused by fragrance compounds, it is essential to assess their eco-toxicological impact, adhering to regulatory compliance. These standard tests provided by OECD (https://www.oecd.org/en/topics/sub-issues/testing-of-chemicals/test-guidelines.html) are performed across different species and ecosystems, evaluating toxicity, persistence, and bioaccumulation of fragrance compounds, thereby determining lethal, sublethal, and chronic effects on various species (Table 4.4). Further, bioaccumulation testing measures the tendency of compound to accumulate in different species and move through the food chain. To determine the persistence potential of fragrance compound, tests are performed to evaluate whether the compound degrades naturally in the environment or resists breakdown. Further understanding of ecosystem impacts is critical to calling the chemical safe. For environmental impact assessment, studies analyze how a compound interacts with species at different trophic levels (Salvito et al., 2011).

These eco-toxicological tests are based on advanced analytical techniques for detecting and quantifying fragrance compounds (Schuijt et al., 2021). For instance, gas chromatography-mass spectrometry (GC-MS) and HPLC can identify and quantify fragrance compounds and measure their degradation products and metabolites in water, sediments, and tissues. Similarly, biomarker analysis can be employed for biological impact studies by measuring oxidative stress markers, enzyme activities, or DNA damage in exposed organisms, and fluorescence microscopy can assess structural damage in algae and phytoplankton studies.

4.7 REGULATORY FRAMEWORK FOR SUSTAINABILITY

Several strategies and regulations are being adopted to mitigate biodegradability and eco-toxicological risks of fragrances. Eco-toxicological testing of fragrance compounds is guided by rules and frameworks implemented

Table 4.4 Common toxicity test methods

	Method	*Parameters measured*	*Test species*	*Duration/ Remark*
Aquatic toxicity testing				
Fish toxicity testing	OECD Test Guideline 203 (Acute Fish Toxicity Test)	LC50 Lethal concentration at which 50% of the test population dies	Zebrafish (*Danio rerio*), fathead minnow (*Pimephales promelas*), and rainbow trout (*Oncorhynchus mykiss*)	96 hours for acute toxicity tests
Daphnia toxicity testing	OECD Test Guideline 202 (Daphnia sp Acute Immobilization Test)	EC50 concentration causing 50% immobilization or adverse effects	Water flea (*Daphnia magna*)	Sensitive indicator species for assessing aquatic toxicity
Algal growth inhibition test	OECD Test Guideline 201 (Freshwater Algae and Cyanobacteria, Growth Inhibition Test)	EC50 concentration causing a 50% reduction in growth rate changes in chlorophyll content	*Pseudokirchneriella subcapitata* (green algae)	Exposure for 72 hours; two response variables: average specific growth rate, and yield are measured
Sediment testing	OECD TG 218 (Sediment-Water Chironomid Toxicity Using Spiked Sediment)	Assesses the impact of fragrance compounds that settle in sediments	*Chironomus riparius* (midge larvae)	28 days, reduction in emergence or larval survival or growth is assessed at the end
Terrestrial toxicity testing: Fragrance compounds can enter terrestrial ecosystems through sludge from wastewater treatment plants or improper disposal.				
Soil microbial activity tests	OECD Test Guideline 216 (Soil Microorganisms Nitrogen Transformation Test)	Changes in microbial respiration and nitrogen transformation rates		At least 28 days; tests the effects on essential soil processes like decomposition and nutrient cycling

Earthworm toxicity testing	OECD Test Guideline 207 (Earthworm Acute Toxicity Test)	LC50 lethal concentration at which 50% of earthworms die	*Eisenia fetida*	Key soil organisms and indicators of soil health
Plant growth testing	OECD TG Test No. 208: Terrestrial Plant Test: Seedling Emergence and Seedling Growth Test	Evaluates the impact on germination rates, biomass, and growth inhibition	-	Evaluated for effects following 14–21 days after 50% emergence of the seedlings in the control group
Bioaccumulation testing: Some fragrance compounds, especially synthetic musks, have high bioaccumulation potential. These tests evaluate whether these substances accumulate in organisms and biomagnify through the food chain				
Fish bioaccumulation test	OECD Test Guideline 305 (Bioconcentration Flow-Through Fish Test)	BCF (Bioconcentration Factor): The ratio of the chemical concentration in the fish to that in the surrounding water	Typically, fish like rainbow trout (*Oncorhynchus mykiss*)	Done in two phases: the exposure (uptake) (28 days normally and 60 days maximum) and post-exposure (depuration) phases
Invertebrate bioaccumulation test		Accumulation of fragrance compounds in species like mussels or crustaceans		
Degradability and persistence testing: Testing the biodegradability and persistence of fragrance compounds is essential for understanding their environmental fate.				
Ready biodegradability test	OECD Test Guideline 301 (Ready Biodegradability)	Percentage of compound mineralized into CO_2 Tests include dissolved oxygen consumption and CO_2 evolution		28 days

(Continued)

Table 4.4 (Continued) Common toxicity test methods

	Method	*Parameters measured*	*Test species*	*Duration/ Remark*
Simulation tests in environmental media	OECD Test Guideline 308 (Aerobic and Anaerobic Transformation in Aquatic Sediment Systems)	Simulates degradation in sediments under both aerobic and anaerobic conditions	At least two sediments different with respect to organic carbon content and texture are used	Maximum 100 days, and should continue until the degradation pathway and water/sediment distribution pattern are established or when 90% of the test substance has been removed by transformation and/or volatilization
Half-life estimation	Evaluates the time it takes for the concentration of the compound to decrease by 50%			
Endocrine disruption testing: Many fragrance compounds, especially synthetic musks, have endocrine-disrupting properties.				
Fish reproductive toxicity test	OECD Test Guideline 229 (Fish Short-Term Reproduction Test)	Changes in reproductive output, hormonal levels, and gonad development	Performed in fathead minnow *(Pimephales promelas)*	21-day exposure period
Vitellogenin assay	Juvenile Medaka anti-Androgen Screening Assay	Measures vitellogenin levels (a biomarker for estrogenic activity)	Model used medaka (*Oryzias latipes*)	28 days (pre-exposure period is not required)

Note: Testing is typically conducted following internationally recognized guidelines, such as those from the OECD (Organization for Economic Co-operation and Development), EPA (Environmental Protection Agency), and ISO (International Organization for Standardization).

Table 4.5 Regulatory standards for biodegradability

Agency	*Role*	*Requirement*
Registration, Evaluation, Authorisation and Restriction of Chemicals (REACH) European Union (EU)	• Mandates comprehensive biodegradability data for all chemical substances sold in the EU. • Substances classified as *Persistent, Bioaccumulative, and Toxic (PBT)* or *very Persistent and very Bioaccumulative (vPvB)* are restricted or banned. • The European Union's REACH regulation mandates biodegradability assessments for chemicals, including fragrance compounds, to limit the release of persistent pollutants.	Fragrance ingredients that fail biodegradability tests or exhibit eco-toxic effects are subject to regulatory limits.
EU Detergents Regulation (EC No 648/2004)	Mandates that surfactants and fragrance ingredients in cleaning and cosmetic products must meet specific biodegradability standards to minimize environmental harm.	
United States Environmental Protection Agency (US EPA)	Enforces the Toxic Substances Control Act (TSCA), which requires biodegradability data for industrial and consumer chemicals, including fragrances.	Substances failing biodegradability or exhibiting high toxicity are flagged for restriction or reformulation.
International Fragrance Association (IFRA) Standards	• IFRA sets safety and environmental guidelines for fragrance use in consumer products, including cosmetics, which include considerations for environmental impact. • Encourages the development and use of readily biodegradable fragrance molecules. • Sets limits or bans on specific fragrance compounds that are known to persist in the environment (e.g., nitro musks). • While IFRA primarily focuses on human safety, these guidelines also consider the environmental burden of fragrance ingredients.	

globally by different governing bodies of different countries (Table 4.5). European (EU) cosmetics guidelines regulate the safety of fragrances to minimize environmental damage and limit the use of certain harmful compounds. Registration, Evaluation, Authorisation and Restriction of Chemicals (REACH) is an EU regulation that requires data on PBT for chemicals following the guidelines in the guidance document (https://eur-lex.europa.eu/

eli/reg/2009/1223/oj/eng). United States Environmental Protection Agency (US EPA) implemented the Toxic Substances Control Act (TSCA) to ensure eco-toxicological evaluations of fragrances (https://www.epa.gov/laws-regulations/summary-toxic-substances-control-act). Additionally, there are International Fragrance Association (IFRA) guidelines, which provide a set of standards for testing fragrances, ensuring their safe use in consumer products. The guidelines focus on the safety assessment of all fragrance ingredients, curtailing harmful health and environmental effects. Therefore, certain harmful ingredients are restricted or limited, promoting safer alternatives (https://ifrafragrance.org/safe-use/library).

Since fragrances are a vital part of cosmetic or personal care products, their biodegradability and eco-toxicological impacts are of serious concern. Due to the source of origin and chemical composition, the biodegradability significantly varies between natural and synthetic fragrance compounds. Natural fragrances, especially plant-based essential oils, are generally more biodegradable but may still pose risks in high concentrations or due to specific compounds, like sesquiterpenes from tree oils. On the other hand, synthetic fragrances like musks or phthalates are less biodegradable, thus contributing to long-term environmental pollution. Many synthetic fragrances persist in the environment and pose a risk to aquatic ecosystems, wildlife, and even human health due to bioaccumulation and endocrine disruption. Due to the harmful effects of certain fragrance compounds and their environmental concerns, it is imperative to improve cosmetic formulation and look for biodegradable alternatives. Sustainable development for fragrances and regulatory initiatives can help alleviate these issues by promoting biodegradable and eco-friendly alternatives.

4.8 CHALLENGES AND PROSPECTS

Fragrance compounds pose significant risks to the ecosystem as these compounds vary in persistence, toxicity, and biodegradability. Although these compounds are fundamental ingredients of consumer product formulations, they are widely reported and distributed across aquatic and terrestrial ecosystems. This highlights the need to address the biodegradation and eco-toxicological challenges such compounds may pose. Interestingly, risk assessment of fragrance compounds to mitigate environmental risks is usually performed as per international regulations and guidelines. Many countries, including the EU, the US, and others, instituted international guidelines (TSCA, REACH, IFRA) to focus on safe development, restrict the use of PBT substances in fragrances, and set limits on eco-toxic fragrance compounds to reduce environmental harm. Adhering to these guidelines, eco-toxicological testing is done to assess their effects on aquatic and terrestrial models, biodegradability, and bioaccumulation potential over a period of time. Despite regulations, there are gaps in understanding the long-term impacts of such compounds on the environment and human health. Many studies focus on acute toxicity during safety and efficacy testing, limiting the understanding of long-term and sub-lethal impacts. Many compounds have been used lately, but residue profiling is not reported often due to the limitation of detection methods, which usually limits the studies on the biodegradation pathways of microbes in the real environment. Thus, studies on biodegradation pathways and intermediate by-products of these fragrance compounds are essential for understanding their ecological safety and sustainability. Another limitation in profiling the residues may be attributed to either the mixture complexity of these fragrance compounds or the occurrence in trace amounts. As discussed in Section 4.3, standard methods exist to perform the toxicity assessments outlined by OECD protocols. However, such guidelines often overlook the effects of mixture complexity and need attention as real-world exposure involves the combined effect of complex mixtures of fragrance compounds and could not be quantified. Also, detecting low concentration would require sensitive methods, which can aid in assessing its eco-toxicity impact.

To address such challenges, in silico testing needs to be developed based on computational models to predict the eco-toxicological effects based on chemical structures, and the combined effect of such chemical structures can be predicted to evaluate and address the risk assessment. Additionally, high-throughput screening methods can address risk assessments in a fast-paced manner using automated systems for multiple endpoints.

Besides, an industrial focus on eco-friendly fragrance alternatives can address the upcoming ecological impact challenges with fragrance compounds. Green chemistry-based solutions are needed for a sustainable environment wherein the development of non-toxic, biodegradable fragrance molecules can be prioritized. Also, persistent and bio-accumulative substances can be excluded from product development involving reformulations. Renewable biodegradable sources yielding biosynthetic or plant-derived terpenes can be exploited to develop safer molecules for organisms and the environment. Biotech-derived fragrances (produced using microorganisms or enzymes) offer a promising alternative. Another method can be employing advances in green chemistry principles to design fragrance molecules that degrade rapidly and safely.

In continuation of sustainable product development, newer testing methods that are faster and more accurate in predicting biodegradability and eco-toxicity without extensive laboratory testing are anticipated. Additionally, as environmental and health concerns of fragrance products grow, biodegradability standards must be revisited, and stricter regulations related to biodegradability and safety limits on fragrance ingredients can be developed and implemented. Compliance with standards is often a significant challenge; thus, a global harmonization of biodegradability criteria and testing protocols across countries will simplify compliance for reasonable product development meeting the sustainability and safety goals.

REFERENCES

Aalizadeh, R., von der Ohe, P. C., & Thomaidis, N. S. (2017). Prediction of acute toxicity of emerging contaminants on the water flea *Daphnia magna* by ant colony optimization–support vector machine QSTR models. *Environ. Sci. Process. Impacts*. 19:438–448. doi: 10.1039/C6EM00679E.

Acharya, K., Werner, D., Dolfing, J., Barycki, M., Meynet, P., Mrozik, W., Komolafe, O., Puzyn, T., & Davenport, R. J. (2019). A quantitative structure-biodegradation relationship (QSBR) approach to predict biodegradation rates of aromatic chemicals. *Water Res*. 157:181–190. doi: 10.1016/j.watres.2019.03.086.

Adamczyk, S., Kiikkilä, O., Kitunen, V., & Smolander, A. (2013). Potential response of soil processes to diterpenes, triterpenes and tannins: Nitrification, growth of microorganisms and precipitation of proteins. *Appl. Soil Ecol*. 67:47–52. doi: 10.1016/J.APSOIL.2013.02.009.

Al-Saleh, I., Al-Rajudi, T., Al-Qudaihi, G., & Manogaran, P. (2017). Evaluating the potential genotoxicity of phthalates esters (PAEs) in perfumes using in vitro assays. *Environ. Sci. Pollut. Res*. 24:23903–23914. doi: 10.1007/S11356-017-9978-1.

Api, A. M., Belsito, D., Bhatia, S., Bruze, M., Calow, P., & Dagli, M. L. et al (2015). RIFM fragrance ingredient safety assessment, benzyl acetate, CAS registry number 140-11-4. *Food Chem. Toxicol*. 84:S15–S24. doi: 10.1016/j.fct.2015.07.010.

Api, A. M., Belsito, D., Botelho, D., Bruze, M., Burton, G. A., & Buschmann, J., et al. (2021a). RIFM fragrance ingredient safety assessment, p-cresol, CAS registry number 106-44-5. *Food Chem. Toxicol*. 158. doi: 10.1016/j.fct.2021.112631.

Api, A. M., Belsito, D., Botelho, D., Bruze, M., Burton, G. A., & Buschmann, J. et al (2021b). RIFM fragrance ingredient safety assessment, phenoxyacetaldehyde, CAS registry number 2120-70-9. *Food Chem. Toxicol*. 156:112505. doi: 10.1016/j.fct.2021.112505.

Api, A. M., Belsito, D., Botelho, D., Bruze, M., Burton, G. A., & Cancellieri, M. A., et al. (2022). RIFM fragrance ingredient safety assessment, tetrahydro-2-isobutyl-4-methyl-2H-pyrane, CAS registry number 13477-62-8. *Food Chem. Toxicol*. 167. doi: 10.1016/j.fct.2022.113323.

Arruda, V., Simões, M., & Gomes, I. B. (2022). Synthetic musk fragrances in water systems and their impact on microbial communities. *Water*. 14(5):692. doi: 10.3390/w14050692.

Ashcroft, S., Dosoky, N. S., Setzer, W. N., & Satyal, P. (2024). Synthetic endocrine disruptors in fragranced products. *Endocrines*. 5:366–381. doi: 10.3390/endocrines5030027.

Azaroff, A., Monperrus, M., Miossec, C., Gassie, C., & Guyoneaud, R. (2021). Microbial degradation of hydrophobic emerging contaminants from marine sediment slurries (Capbreton Canyon) to pure bacterial strain. *J. Hazard Mater*. 402:123477. doi: 10.1016/j.jhazmat.2020.123477.

Bicas, J. L., Dionísio, A. P., & Pastore, G. M. (2009). Bio-oxidation of terpenes: An approach for the flavor industry. *Chem. Rev*. 109(9): 4518–4531. doi:10.1021/cr800190y

Broccoli, A., Anselmi, S., Cavallo, A., Ferrari, V., Prevedelli, D., Pastorino, P., & Renzi, M. (2021). Ecotoxicological effects of new generation pollutants (nanoparticles, amoxicillin and white musk) on freshwater and marine phytoplankton species. *Chemosphere*. 279:130623. doi: 10.1016/j.chemosphere.2021.130623.

Cadwallader, K. R., Braddock, R. J., Parish, M. E., & Higgins, D. P. (1989). Bioconversion of (+)-limonene by *Pseudomonas gladioli*. *J. Food Sci*. 54(5): 1241–1245. doi:10.1111/j.1365-2621.1989.tb05964.x.

Canterino, M., Marotta, R., Temussi, F., & Zarrelli, A. (2008). Photochemical behaviour of musk tibetene. A chemical and kinetic investigation. *Environ. Sci. Pollut. Res. Int.* 15(3):182–187. doi: 10.1065/espr2007.12.464.

Chane, A. D., Košnář, Z., Hřebečková, T., Wiesnerová, L., Jozífek, M., Doležal, P., Praus, L., & Tlustoš, P. (2023). Bioremediation of the synthetic musk compounds galaxolide and tonalide by white rot fungal strain-assisted phytoremediation in biosolid-amended soil. *Chemosphere.* 328:138605. doi: 10.1016/j.chemosphere.2023.138605.

Chatterjee, T., & Bhattacharyya, D. K. (2001). Biotransformation of limonene by *Pseudomonas putida. Appl. Microbiol. Biotechnol.* 55(5):541–546. doi: 10.1007/s002530000538.

Clara, M., Gans, O., Windhofer, G., Krenn, U., Hartl, W., Braun, K., Scharf, S., & Scheffknecht, C. (2011). Occurrence of polycyclic musks in wastewater and receiving water bodies and fate during wastewater treatment. *Chemosphere.* 82(8):1116–1123. doi: 10.1016/j.chemosphere.2010.11.041.

Dick, C., Rey, S., Boschung, A., Miffon, F., & Seyfried, M. (2016). Current limitations of biodegradation screening tests and prediction of biodegradability: A focus on fragrance substances. *Environ. Technol. Innov.* 5:208–224. doi: 10.1016/J.ETI.2016.03.002.

Duan, W., Meng, F., Cui, H., Lin, Y., Wang, G., & Wu, J. (2018). Ecotoxicity of phenol and cresols to aquatic organisms: A review. *Ecotoxicol. Environ. Saf.* 157:441–456. doi: 10.1016/J.ECOENV.2018.03.089.

Duetz, W. A., Fjällman, A. H., Ren, S., Jourdat, C., & Witholt, B. (2001). Biotransformation of d-limonene to (+) trans-carveol by toluene-grown *Rhodococcus opacus* PWD4 cells. *Appl. Environ. Microbiol.* 67(6):2829–2832. doi: 10.1128/AEM.67.6.2829-2832.2001.

ECB (2005a). European Union Risk Assessment Report: 5-tert-butyl-2,4,6-trinitro-m-xylene (musk xylene). European Chemicals Bureau, Bilthoven, The Netherlands. http://echa.europa.eu.

ECB (2005b). European Union Risk Assessment Report: 4′-tert-butyl-2′,6′-dimethyl-3′,5′-dinitroacetophenone (musk ketone). European Chemicals Bureau, Bilthoven, The Netherlands. http://echa.europa.eu.

Gao, Y., Li, G., Qin, Y., Ji, Y., Mai, B., & An, T. (2019). New theoretical insight into indirect photochemical transformation of fragrance nitro-musks: Mechanisms, eco-toxicity and health effects. *Environ. Int.* 129: 68–75. doi:10.1016/j.envint.2019.05.020.

Gooding, M. P., Newton, T. J., Bartsch, M. R., & Hornbuckle, K. C. (2006). Toxicity of synthetic musks to early life stages of the freshwater mussel *Lampsilis cardium. Arch. Environ. Contam. Toxicol.* 51:549. doi: 10.1007/S00244-005-0223-4.

Gueretz, J. S., Somensi, C. A., Martins, M. L., & Souza, A. P. (2017). Evaluation of eugenol toxicity in bioassays with test-organisms. *Cienc. Rural.* 47:e20170194. doi: 10.1590/0103-8478cr20170194.

Herren, D., & Berset, J. D. (2000). Nitro musks, nitro musk amino metabolites and polycyclic musks in sewage sludges: Quantitative determination by HRGC-ion-trap-MS/MS and mass spectral characterization of the amino metabolites. *Chemosphere.* 40(5):565–574. doi:10.1016/S0045-6535(99)00325-2

Jin, K. S., Bak, M. J., Jun, M., Lim, H. J., Jo, W. K., & Jeong, W. S. (2010). α-pinene triggers oxidative stress and related signaling pathways in A549 and HepG2 cells. *Food Sci. Biotechnol.* 19:1325–1332. doi: 10.1007/S10068-010-0189-5.

Jing, Y., Sijia, H., Sijia, L., Fuxiang, T., Xiyan, J., Yuanting, L., Meifang, H., Wenwu, X., Liang, M., & Lijuan, S. (2022). Enantioselective effects of chiral fragrance carvone (L- and D-carvone) on the physiology, oxidative damage, synthesis, and release of microcystin-LR in *Microcystis aeruginosa. Sci. Total Environ.* 853:158631. doi: 10.1016/j.scitotenv.2022.158631.

Käfferlein, H. U., Göen, T., & Angerer, J. (1998). Musk xylene: Analysis, occurrence, kinetics, and toxicology. *Crit. Rev. Toxicol.* *28*(5): 431–476. doi:10.1080/10408449891344245

Kempińska-Kupczyk, D., & Kot-Wasik, A. (2019). The potential of LC–MS technique in direct analysis of perfume content. *Monatshefte fur Chemie.* 150:1617–1623. doi: 10.1007/S00706-019-02468-6.

Klaschka, U., & Kolossa-Gehring, M. (2007). Fragrances in the environment: Pleasant odours for nature? (9 pp). *Environ. Sci. Pollut. Res. Int.* 14(Suppl 1):44–52. doi: 10.1065/ESPR2007.01.380.

Košnář, Z., Mercl, F., Chane, A. D., Pierdonà, L., Míchal, P., & Tlustoš, P. (2021). Occurrence of synthetic polycyclic and nitro musk compounds in sewage sludge from municipal wastewater treatment plants. *Sci. Total Environ.* 20; 801:149777. doi: 10.1016/j.scitotenv.2021.149777.

Li, X., Chu, Z., Yang, J., Li, M., Du, M., & Zhao, X. et al (2018). Synthetic musks: A class of commercial fragrance additives in personal care products (PCPs) causing concern as emerging contaminants. *Adv. Mar. Biol.* 81:213–280. doi: 10.1016/BS.AMB.2018.09.008.

Lin, J., & Emberger, M. (2017). Study of the photodegradation of a fragrance ingredient for aquatic environmental fate assessment. *Chemosphere.* 173:485–493. doi: 10.1016/j.chemosphere.2017.01.083.

Lotfi, S., Ahmadi, S., & Kumar, P. (2022). Ecotoxicological prediction of organic chemicals toward *Pseudokirchneriella subcapitata* by Monte Carlo approach. *RSC Adv.* 12:24988. doi: 10.1039/D2RA03936B.

MacKintosh, R. W., & Fewson, C. A. (1988). Benzyl alcohol dehydrogenase and benzaldehyde dehydrogenase II from *Acinetobacter calcoaceticus*. Substrate specificities and inhibition studies. *Biochem. J.* 255(2):653–661.

Martin, C., Moeder, M., Daniel, X., Krauss, G., & Schlosser, D. (2007). Biotransformation of the polycyclic musks HHCB and AHTN and metabolite formation by fungi occurring in freshwater environments. *Environ. Sci. Technol.* 41(15):5395–5402. doi: 10.1021/es0711462.

Molina, G., Pessôa, M. G., Bicas, J. L., Fontanille, P., Larroche, C., & Pastore, G. M. (2019). Optimization of limonene biotransformation for the production of bulk amounts of α-terpineol. *Bioresour. Technol.* 294: 122180. doi:10.1016/j.biortech.2019.122180.

Musilova, L., Ridl, J., Polivkova, M., Macek, T., & Uhlik, O. (2016). Effects of secondary plant metabolites on microbial populations: Changes in community structure and metabolic activity in contaminated environments. *Int. J. Mol. Sci.* 17:1205. doi: 10.3390/ijms17081205.

Nazzaro, F., Fratianni, F., De Martino, L., Coppola, R., & De Feo, V. (2013). Effect of essential oils on pathogenic bacteria. *Pharm.* 6:1451–1474. doi: 10.3390/ph6121451.

OECD (1992). Test No. 301: Ready Biodegradability OECD Guidelines for the Testing of Chemicals, Section 3, OECD Publishing, Paris, doi:10.1787/9789264070349-EN.

OECD (2009). *Test No. 302C*: Inherent Biodegradability: Modified MITI Test (II) OECD Guidelines for the Testing of Chemicals, Section 3, OECD Publishing, Paris. doi:10.1787/9789264070400-EN.

OECD (2014). Test No. 310: Ready Biodegradability - CO2 in sealed vessels (Headspace Test), OECD Guidelines for the Testing of Chemicals, Section 3, OECD Publishing, Paris. doi: 10.1787/9789264224506-en.

Osemwengie, L. I., & Steinberg, S. (2001). On-site solid-phase extraction and laboratory analysis of ultra-trace synthetic musks in municipal sewage effluent using gas chromatography-mass spectrometry in the full-scan mode. *J. Chromatogr. A.* 932: 107–118. doi: 10.1016/s0021-9673(01)01216-x.

OSPAR Commission (2004). OSPAR background document on musk xylene and other musks. The Convention for the Protection of the Marine Environment of the North-East Atlantic (OSPAR Convention) Commission, OSPAR publication. https://www.ospar.org/documents?v=6978.

Pastor-Nieto, M. A., & Gatica-Ortega, M. E. (2021). Ubiquity, hazardous effects, and risk assessment of fragrances in consumer products. *Curr. Treat. Options Allergy.* 8:21. doi: 10.1007/S40521-020-00275-7.

Priefert, H., Overhage, J., & Steinbüchel, A. (1999). Identification and molecular characterization of the eugenol hydroxylase genes (ehyA/ehyB) of *Pseudomonas* sp. strain HR199. *Arch. Microbiol.* 172:354–363.

Priefert, H., Rabenhorst, J., & Steinbüchel, A. (1997). Molecular characterization of genes of *Pseudomonas* sp. strain HR199 involved in bioconversion of vanillin to protocatechuate. *J Bacteriol.* 179(8):2595–2607. doi: 10.1128/jb.179.8.2595-2607.1997.

Ravichandran, C., Badgujar, P. C., Gundev, P., & Upadhyay, A. (2018). Review of toxicological assessment of d-limonene, a food and cosmetics additive. *Food Chem. Toxicol.* 120:668–680. doi: 10.1016/J.FCT.2018.07.052.

Rimkus, G. G., Gatermann, R., & Hühnerfuss, H. (1999). Musk xylene and musk ketone amino metabolites in the aquatic environment. *Toxicology letters.* 111(1-2): 5–15. doi: 10.1016/s0378-4274(99)00190-3.

Rönnander, J., Ljunggren, J., Hedenström, E., & Wright, S. A. I. (2018). Biotransformation of vanillin into vanillyl alcohol by a novel strain of *Cystobasidium laryngis* isolated from decaying wood. *AMB Express.* 8(1):137. doi: 10.1186/s13568-018-0666-4.

Salvito, D., Lapczynski, A., Sachse-Vasquez, C., McIntosh, C., Calow, P., & Greim, H. et al (2011). Macrocyclic fragrance materials—A screening-level environmental assessment using chemical categorization. *Ecotoxicol. Environ. Saf.* 74:1619–1629. doi: 10.1016/J.ECOENV.2011.05.002.

Schuijt, L. M., Peng, F. J., van den Berg, S. J. P., Dingemans, M. M. L., & Van den Brink, P. J. (2021). (Eco)toxicological tests for assessing impacts of chemical stress to aquatic ecosystems: Facts, challenges, and future. *Sci. Total Environ.* 795:148776. doi: 10.1016/J.SCITOTENV.2021.148776.

Singh, A. K., Bilal, M., Iqbal, H. M. N., & Raj, A. (2021). Trends in predictive biodegradation for sustainable mitigation of environmental pollutants: Recent progress and future outlook. *Sci. Total Environ.* 770:144561. doi: 10.1016/J.SCITOTENV.2020.144561.

Tas, J. W., Balk, F., Ford, R. A., & Van de Plassche, E. J. (1997). Environmental risk assessment of musk ketone and musk xylene in the Netherlands in accordance with the EU-TGD. *Chemosphere.* 35(12): 2973–3002. doi: 10.1016/S0045-6535(97)00261-0.

Tasseli, S., Rogora, M., Orrù, A., & Guzzella, L. (2023). Behaviour of synthetic musk fragrances in freshwaters: Occurrence, relations with environmental parameters, and preliminary risk assessment. *Environ. Sci. Pollut. Res. Int.* 30(50): 109643–109658. doi:10.1007/s11356-023-30030-9

Tasselli, S., Valenti, E. & Guzzella, L. (2021). Polycyclic musk fragrance (PMF) removal, adsorption and biodegradation in a conventional activated sludge wastewater treatment plant in Northern Italy. *Environ Sci Pollut Res.* 28: 38054–38064. doi: 10.1007/s11356-021-13433-4.

US EPA (2008). Estimations Programs Interface (EPI) Suite for Microsoft Windows®, v 4.10. United States Environmental Protection Agency, Washington DC. Available at http://www.epa.gov.

US EPA (2012). The ECOSAR (ECOlogical Structure Activity Relationship) Class Program for Microsoft Windows®, v 1.11. United States Environmental Protection Agency, Washington DC. Available at http://www.epa.gov.

van der Werf, M. J., Swarts, H. J., & de Bont, J. A. M. (1999). *Rhodococcus erythropolis* DCL14 contains a novel degradation pathway for limonene. *Appl. Environ. Microbiol.* 65: 2092–2102. doi:10.1128/AEM.65.5.2092-2102.1999.

Wilkinson, J. L., Hooda, P. S., Barker, J., Barton, S., & Swinden, J. (2016). Ecotoxic pharmaceuticals, personal care products, and other emerging contaminants: A review of environmental, receptor-mediated, developmental, and epigenetic toxicity with discussion of proposed toxicity to humans. *Crit. Rev. Environ. Sci. Technol.* 46:336–381. doi: 10.1080/10643389.2015.1096876.

Zhang, X., Brar, S. K., Yan, S., Tyagi, R. D., & Surampalli, R. Y. (2013a). Fate and transport of fragrance materials in principal environmental sinks. *Chemosphere.* 93:857–869. doi: 10.1016/j.chemosphere.2013.05.055.

Zhang, Y., Jiao, Y., Li, Z., Tao, Y., & Yang, Y. (2021). Hazards of phthalates (PAEs) exposure: A review of aquatic animal toxicology studies. *Sci. Total Environ.* 771:145418. doi: 10.1016/j.scitotenv.2021.145418.

Zhang, X., Xu, Q., Man, S., Zeng, X., Yu, Y., & Pang, Y., et al. (2013b). Tissue concentrations, bioaccumulation, and biomagnification of synthetic musks in freshwater fish from Taihu Lake, China. *Environ. Sci. Pollut. Res.* 20: 311–322. doi:10.1007/S11356-012-1095-6.

Chapter 5

Good manufacturing practices (GMP) in production of essential oils/fragrances for cosmetics industry

Saroj Kumar Amar
Department of Forensic Science, School of Basic and Applied Sciences, Adamas University, Kolkata, India

Ayushi Mishra
Department of Biochemistry, University of Lucknow, Lucknow, India

Santosh Kumar Shukla
Corporate R&D Centre, HLL Lifecare Ltd., Thiruvananthapuram, India

5.1 INTRODUCTION

Essential oils are extensively utilized in the cosmetics and food sectors for their aromatic and flavoring properties. Additionally, they serve as active pharmaceutical ingredients (APIs) in the production of herbal medicinal products, classifying them as "atypical actives." Consequently, while the good manufacturing practices (GMP) guidelines outlined in Part II of the European Guidelines to GMP are generally applicable, these same essential oils are often manufactured under alternative standards. For such "atypical actives," a pragmatic, risk-based approach, evaluated on a case-by-case basis, is essential to guarantee product quality and safety. Annex 7 of the European Guideline to GMP on the production of herbal medicinal products already describes a pragmatic approach for the initial stages of essential oil manufacture. However, ongoing discussions to amend Directive 2001/83/EC concerning falsified medicines include stricter requirements for APIs, such as mandatory GMP certificates, which could pose challenges for essential oil producers, particularly in the initial processing steps.

GMP are a fundamental set of guidelines and procedures designed to ensure the consistent production and quality control (QC) of products, especially essential oils and fragrances used in cosmetics. These standards aim to minimize inherent production risks that could compromise product quality and safety. For the essential oils and fragrances industry, GMP act as a crucial safeguard against issues like contamination, mislabeling, and inconsistencies in product quality. The application of GMP is particularly vital due to the complex handling requirements of volatile and sensitive ingredients like essential oils, which demand meticulous attention during manufacturing and storage.

DOI: 10.1201/9781003558477-5

5.2 GMP GUIDELINES IN THE COSMETICS INDUSTRY

The implementation of GMP within the cosmetics industry involves a comprehensive framework of guidelines to assure product safety, quality, and consistency. These guidelines are derived from globally recognized standards established by organizations such as the World Health Organization (WHO) and the International Organization for Standardization (ISO 22716), as well as region-specific regulations like the FDA's Voluntary Cosmetic Registration Program in the United States and the Cosmetic Regulation (EC) No. 1223/2009 in the European Union.

5.2.1 Personnel and training

All personnel involved in manufacturing must receive thorough training in hygiene, sanitation, and specific production processes. This training is crucial for preventing contamination and ensuring the correct handling of sensitive ingredients such as essential oils, which can be easily compromised by improper handling. Clearly defined responsibilities for each staff member and regular refresher training programs are necessary to maintain up-to-date knowledge of regulations and best practices. Cosmetic GMP emphasize the importance of adequately trained, experienced, and qualified employees whose conduct supports safety and quality. The company must provide the necessary support, including training, supervision, education, resources, and safety equipment, to enable staff to perform their duties cleanly and appropriately.

5.2.2 Premises and equipment

Manufacturing facilities must be maintained in a clean, well-lit, and properly ventilated condition. Equipment used in production should undergo regular maintenance and calibration to prevent contamination and ensure accurate ingredient measurement and mixing. Environmental controls are particularly important in areas where essential oils are blended, as these volatile compounds are susceptible to degradation or contamination from temperature, light, and air exposure. All equipment, including containers, utensils, and measuring tools, must be clean and designed to prevent the adulteration of finished cosmetics with foreign materials, dirt, or cleaning residues. Proper storage and periodic calibration are required, and unsuitable or broken equipment should be promptly repaired or replaced.

5.2.3 Raw materials and supplier management

All raw materials must be systematically organized and clearly labeled with essential information (name, lot number, and control status) to ensure traceability throughout the manufacturing process. Specific guidelines address water quality, color additives, and restricted cosmetic ingredients. Traceability of the initial plant raw material to its use in cosmetic products is mandatory. This can be achieved through supplier licenses (if applicable), documented supplier commitments, or through techniques such as:

- *Macroscopic botanical characteristics*: Comparing the plant material to a reference description for rapid identification.
- *Microscopic botanical characteristics*: Examining the plant microscopically to identify specific features and detect foreign elements.
- *Thin layer chromatography (TLC) or gas chromatography (GC)*: Comparing the chromatogram of the test solution with a control solution containing reference substances (TLC) or a chromatographic profile (GC).

GMP mandate that all raw materials, including essential oils, fragrances, and other cosmetic ingredients, be sourced from reputable suppliers. The quality and authenticity of these materials must be verified through testing before use. Proper labeling, storage in appropriate conditions (e.g., cool, dark environments for essential oils and fragrances to maintain stability), and robust traceability systems are essential for all raw materials.

5.2.4 Production process control

Comprehensive documentation and records of all production aspects are required. Establishing standard operating procedures (SOPs) for all processes is recommended. An SOP is a detailed written instruction outlining how a specific production activity is to be conducted. Clear procedures must be defined for all stages of production, including precise instructions for mixing, heating, cooling, and packaging to ensure batch-to-batch consistency. Critical control points within the process should be identified to facilitate early detection of quality deviations. Regular in-process checks at these critical stages are a key GMP requirement.

5.2.5 Quality control (QC)

QC is a critical element in essential oil manufacturing, ensuring the production of safe, pure, and effective products. Given their concentrated nature and diverse applications, stringent QC measures are necessary throughout the manufacturing process to maintain their therapeutic properties and safety. QC involves rigorous testing of both raw materials and finished products to verify their quality and safety. Each batch must undergo thorough testing to ensure compliance with established specifications. Particular attention should be paid to fragrances and essential oils, which may require specific analyses for purity, potential allergens, or irritants. For example, certain essential oils, like citrus or lavender, can exhibit phototoxic or allergenic properties if not correctly formulated.

5.2.6 Packaging and labeling

Appropriate packaging and storage are vital for preventing contamination and degradation of essential oils, and QC extends to these stages to maintain product integrity. GMP emphasize the use of suitable packaging to ensure product stability. Packaging for cosmetics containing essential oils should protect them from environmental factors such as UV light, oxygen, and moisture that can compromise their quality. Labels must clearly list all ingredients, especially any known allergens or potential irritants, to ensure consumer safety. Regulations, such as those under the EU's Cosmetics Regulation, require the listing of fragrance allergens exceeding specific thresholds.

5.2.7 Documentation and record keeping

Documentation, the act of recording how processes are carried out, results in records that collectively form the body of documentation. Cosmetic GMP stress the importance of documenting all production processes and workflows, whether in paper or electronic format. Beyond documenting procedures, records of what was actually done for each batch are essential. Accurate record-keeping enables businesses to quickly identify the cause of issues when they arise, providing evidence of GMP adherence. Detailed records of all manufacturing steps, including the origin and quality of raw materials, batch numbers, in-process testing results, and final product evaluations, must be maintained. GMP highlight that accurate documentation ensures traceability and accountability. These records are also crucial for facilitating product recalls, if necessary, and for supporting audits conducted by regulatory bodies for GMP certification.

5.3 GMP CERTIFICATION PROCESSES

GMP certification is obtained through audits and inspections conducted by independent, recognized certification bodies. The certification process typically involves the following steps:

5.3.1 Pre-audit assessment

An internal audit is crucial for evaluating the overall performance of the quality management system. It should identify strengths and weaknesses and propose corrective actions to address GMP deficiencies. An effective internal audit is unbiased and aims for continuous improvement through rigorous self-evaluation. Prior to an official GMP audit, the manufacturer must conduct a comprehensive internal review to ensure compliance with all GMP requirements. This self-assessment includes evaluating production processes, QC procedures, personnel training, and documentation practices.

For companies handling essential oils and fragrances, particular attention must be paid to ingredient purity, handling protocols, and potential contamination risks.

5.3.2 Third-party audit

A qualified third-party auditor will visit the manufacturing facility to perform a detailed assessment of the GMP system. During the audit, the auditor verifies compliance with all relevant GMP guidelines, including hygiene practices, process controls, and testing procedures. The audit will specifically examine the handling of sensitive ingredients like essential oils, ensuring that appropriate temperature and light controls are in place to maintain product integrity.

5.3.3 Non-compliance and corrective actions

All complaints or adverse events related to a specific product must be thoroughly recorded and evaluated according to established SOPs. This systematic logging and review process helps businesses identify the root cause of these events, thereby highlighting potential safety or quality issues. For instance, complaints about a product's odor or color may indicate spoilage or contamination, while reports of incorrect labeling could signal a batch-wide mix-up. Adverse events, such as complaints of undesired bodily reactions, require particularly careful evaluation. If any non-compliant areas are identified during the audit, the manufacturer will be given the opportunity to implement corrective actions, which must be documented and presented to the auditor for review.

5.3.4 Issuance of GMP certification

Upon successful completion of the audit and satisfactory resolution of any identified non-compliances, the certification body will issue a GMP certificate. This certification demonstrates that the company adheres to recognized standards for GMP. GMP certification is not a one-time event; regular audits are conducted to ensure ongoing compliance, especially when significant changes are made to the production process or when new ingredients, such as essential oils from different suppliers, are introduced.

5.4 ENSURING PRODUCT QUALITY AND SAFETY WITH GMP

5.4.1 Minimizing contamination risks

GMP guidelines place a strong emphasis on minimizing contamination risks. Essential oils, which are highly susceptible to oxidation and contamination, require meticulous handling and storage. GMP ensure that manufacturers adhere to strict guidelines for cleanroom environments, sterilized equipment, and protective packaging.

5.4.2 Consistency in product quality

Consistency in product quality depends on the ability to control manufacturing variables and processes to achieve a uniform outcome with each batch. Factors such as formulation, raw material selection, sanitation protocols, and the expertise of the cosmetic chemist can influence product quality. Lack of control over these variables can lead to batch-to-batch variations. Accurate and thorough documentation, followed consistently, is vital for reproducing product quality and achieving consistency. Quality itself refers to a product's ability to meet defined criteria based on its attributes and characteristics, often operationalized through objective requirements like color, odor, viscosity, and pH. It also encompasses safety and purity. GMP assist companies in manufacturing products that maintain consistently high quality across all batches. By adhering to GMP guidelines, manufacturers can guarantee that every batch of a cosmetic product meets the same quality and performance standards. This is particularly crucial for cosmetics containing essential oils, where variations in potency and purity can impact the final product's safety and efficacy.

5.4.3 Regulatory compliance

In the European Union, cosmetic products are governed by the Cosmetics Regulation (EC) No. 1223/2009, as amended, with specific provisions in Annex II and Annex III addressing the use of essential oils. This regulation mandates that responsible persons ensure a cosmetic product undergoes a safety assessment based on relevant information before it is placed on the market, requiring the creation of a safety report according to Annex I (with guidelines provided in Commission Implementing Decision 2013/674/EU). Further guidance on cosmetic ingredients is available in Council of Europe publications. GMP assist manufacturers in complying with international regulatory standards such as the EU's Cosmetics Regulation, the FDA's Voluntary Cosmetic Registration Program, and ISO 22716. Meeting these standards is essential for companies seeking to market their products globally.

5.4.4 Consumer safety and trust

Safety in cosmetic production involves preventing unintentional contamination, spoilage, or misuse of final products that could cause adverse reactions and other health effects. This includes practices like sourcing raw materials from reputable suppliers, ensuring proper facility cleaning, educating staff on hygiene, and accurate label proofreading. It is paramount to conduct risk assessments of essential oils in cosmetic products in compliance with Annex I of the Cosmetic Regulation (EC) No. 1223/2009 and related guidelines. This requires a sequential approach based on literature data concerning the essential oil and its chemical components, and toxicological studies to characterize hazards in the absence of sufficient data (as detailed in SCCS guidance). GMP certification assures consumers that the products they use are safe, of high quality, and produced in a controlled environment. This trust is particularly critical in the cosmetics industry, where products like essential oils and fragrances are applied directly to the skin and can pose health risks if improperly formulated.

5.5 CONCLUSION

GMP form a fundamental basis for ensuring product quality, safety, and consistency within the cosmetics industry. For cosmetics incorporating essential oils and fragrances, strict adherence to GMP guidelines is indispensable due to the sensitive nature of these ingredients. By implementing rigorous controls across raw material sourcing, production processes, packaging, and testing, manufacturers can effectively prevent contamination, guarantee product efficacy, and maintain regulatory compliance. Ultimately, GMP not only ensure high standards of product quality but also cultivate consumer trust and enable manufacturers to compete effectively in an increasingly globalized marketplace.

Specific International legislation and recommendations

S. no.	*Country*	*Year*	*Recommendations*
1	France	2008	Recommendations regarding essential oils quality Criteria
2	France	2010	Recommendations for the risk assessment associated with the use of essential oils in cosmetic products
3	France	2008	Recommendations to manufacturers and responsible persons for the placing on the market of cosmetic products containing terpenoids: camphor, eucalyptol, menthol
4	Germany	2008	The Federal Institute for Risk Assessment recommended the following [1]: • Maximum concentration of 1% of camphor, eucalyptus oil, menthol, or methyl salicylate in leave-on products and a maximum concentration of 5% of camphor, 4% of menthol, and 2.5% of methyl salicylate in rinse-off products. • The Federal Institute for Risk Assessment (BfR) has published an opinion on tea tree oil recommending a concentration limit of 1% for tea tree oil in cosmetic products.

(*Continued*)

S. no.	Country	Year	Recommendations
5	Switzerland	2008	The Swiss legislation on cosmetics has fixed a maximal concentration of 3.0% essential oils (alone or in mixture) in leave-on products (Ordinance on cosmetics, SR 817.023.31). This maximal concentration helps to delimitate cosmetics and medical products.
6	Council of Europe	2006	The Council of Europe has published three volumes of recommendations concerning the use of plants and plant-based preparations used as ingredients in cosmetic products, as increasing numbers of plant-based cosmetic products have been placed on the market. The Council of Europe has also published recommendations concerning the use of camphor, eucalyptol, and menthol.

BIBLIOGRAPHY

Bundesinstitut für die Risikobewertung (BfR). (2008). FAQs, 'Frequently Asked Questions about the use of essential oils', 28.02.2008.

Council of Europe. (2008). *Active ingredients used in cosmetics: safety survey.*

Council of Europe. (2001). Plants in cosmetics: plants and herbal preparations used as ingredients in cosmetics.Volume I. Patri F. and Silano V.

Council of Europe. (2001). Plants in cosmetics: plants and herbal preparations used as ingredients in cosmetics. Volume II. Anton R., Patri F. and Silano V.

Council of Europe. (2006). *Plants in cosmetics: potentially harmful components.* Volume III. Committee of Experts on Cosmetic Products.

Council of Europe. *Safe cosmetics for young children. A guide for manufacturers and safety assessors.* ISBN 978-92-871-7337-9.

European Commission. (2009). Cosmetics Regulation (EC) No. 1223/2009.

European Commission. (2009). Regulation 1223/2009 on cosmetic products, as amended, *Official Journal of the European Union*, L342/59-209; 22.12.2009.

European Commission. *The SCCS notes of guidance for the testing of cosmetic ingredients and their safety evaluation*, 8th revision.

International Organization for Standardization (ISO). (2007). ISO 22716:2007 – Cosmetics – Good Manufacturing Practices (GMP) – Guidelines on Good Manufacturing Practices.

US Food and Drug Administration (FDA). Voluntary Cosmetic Registration Program (VCRP). [FDA Website] (https://www.fda.gov/cosmetics/registration-listing/cosmetics-voluntary-registration-program).

World Health Organization (WHO). (2007). Good Manufacturing Practices (GMP): Supplementary guidelines for the manufacture of herbal medicines. WHO Technical Report Series No. 937.

Chapter 6

International regulatory bodies and standards

Overview of key international regulatory organizations, harmonization efforts, and international standards

Shailesh Satish Dixit
Probiotech FNC, Pune, India

6.1 INTRODUCTION

"Kosmeticos" is a Greek word for "cosmetic", which means to embellish. Cosmetics are used for enhancing or maintaining our appearance. It is defined as a "preparation meant for external use on parts of the body (skin, hair, nail, etc.) for cleansing, nourishing, beautifying, colouring, preserving, protecting or removing". It is also defined as any article intended to be rubbed, poured, sprinkled or sprayed on, or introduced into, or otherwise applied to, the human body or any part thereof for cleansing, beautifying, promoting attractiveness, or altering the appearance, and includes any article intended for use as a component of cosmetic [1, 2].

The cosmetics industry has a global footprint and witnessed rapid expansion. It is dynamic in nature with new innovations and the addition of a wide range of new products which is increasing its sales year by year. The global cosmetic market size in 2022 was valued at USD 378 billion and is expected to hit around USD 661.12 billion by 2032 with a compound annual growth rate (CAGR) of 5.8% from 2023 to 2032 [3]. As the cosmetic industry grows, there is an increasing demand for and focus on natural and herbal ingredients-based products.

6.1.1 Essential oils (EOs)

Though essential oils (EOs) are not defined in the cosmetic legislation [4], the more developed definition in terms of application to public health comes from the medicinal product world. The Commission of the European Pharmacopoeia defines EO [5] in a similar way to that of the International Organization for Standardization [ISO/DIS 9235(en)] standards on aromatic natural raw ingredients [6]:

> *Odorous product, usually with a complex composition, obtained from a botanically defined plant raw material by steam distillation, dry distillation, or a suitable mechanical process without heating. Essential oils are usually separated from the aqueous phase by a physical process that does not significantly affect their composition.*

DOI: 10.1201/9781003558477-6

The French National Institute of Standardization, also known as the Agence Francaise de Normalisation (AFNOR), defines EOs as (NF T 75-006): EOs are produced from vegetable raw materials by the processes of steam distillation, mechanical extraction from the citrus epicarp, or "dry" distillation, respectively. After that, the EO is separated from the water with the use of several physical techniques [7, 8].

Increasing interest in natural cosmetics is due to the increasing self-awareness of consumers and the trend of using "eco" influenced products. Today natural cosmetics is a strongly developing segment in the cosmetic industry. The growing willingness of consumers for natural cosmetics is because they are more potent [9] and is reflected by its increased production and sales. Especially cosmeceuticals are in more demand as they contain biologically active ingredients like EOs from plant origin. As natural beauty is a current trend, the popular cosmetic brands opt for natural fragrances or minimally processed natural ingredients. Consumers are paying more attention to the ingredients of cosmetics due to increased awareness and are more attracted to product which has more transparent ingredients [10]. Hence, it becomes important to take care of the environment, product safety, animal welfare, and consumer health.

EOs are used in a variety of products like cosmetics, aromatherapy, flavours, personal hygiene, household items, etc. Hussain et al. [11] have reported that out of about 3,000 plant species, 300 EOs are commercially available in flavours and fragrances market. Few high-valued EOs used as fragrances are lavender, citrus, tea tree, eucalyptus, and other floral oils. Among them geraniol, linalool, citral, limonene, and citronellol are widely appreciated and accepted fragrances in various cosmetics [12]. EOs play an important role in masking the undesirable smell of formulation as well as imparting pleasant aroma, preservative property, and skin benefits due to the presence of active ingredients present in them. EOs are used alone or in combination with other preservatives owing to their antimicrobial properties, which enhance dermato-cosmetic properties of a product, to ensure that the product remains suitable and safe during its shelf-life [13]. Cosmeceutical products induced with EOs have additional benefits to skin (Figure 6.1) like anti-acne, skin lightening, UV protection, anti-aging, etc., thus making them highly valuable to cosmetic industry [12]. Each individual EO has a unique chemical profile and is associated with various benefits. Guzmán et al. [14] have summarized few potential applications of EOs in the design of cosmetic formulations.

6.1.2 Sources of EOs

There are 400,000 known plant species of both aromatic and medicinal plants, out of which about 2,000 species come from nearly 60 botanical families of EOs-bearing plants [12, 15]. Important EOs are found in various taxonomic groups and distributed among all the plant classes [12, 16].

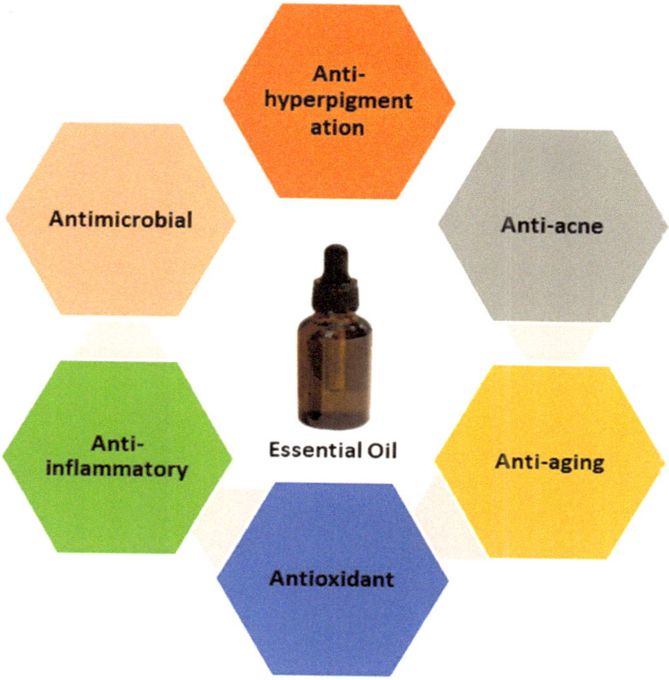

Figure 6.1 Benefits of essential oils.

Parts of plants that serve as major sources of EO can be different for different plants. Like, flowers and inflorescences (e.g., ylang-ylang, rose, lavender, neroli, jasmine, chamomile, hibiscus), leaves (e.g., eucalyptus, tea tree, lemongrass, peppermint, oregano, basil, rosemary), fruits (black pepper, nutmeg), peel (orange, bergamot, grapefruit, lemon, tangerine), seeds (caraway, anise, cumin, cardamom, fennel), berries (mrytle, allspice, juniper), bark (cinnamon, cedar, cassia, sassafras), wood (cedarwood, camphor, sandalwood), root and rhizome (ginger, vetiver, turmeric), and resin (myrrh, frankincense) [17].

6.1.3 Applications of EOs in cosmetic

Nowadays, there is a growing demand for natural cosmetics [13, 18]. The market for natural cosmetics in 2021 reached approximately $40B USD, which is roughly 10% of the global cosmetics and toiletries market [16]. EOs are increasingly used in the production of natural cosmetics due to their active ingredients, fragrances, and preservative properties. Also, they can be used as natural alternatives to synthetic ones commonly found in conventional cosmetics.

EOs are used in various cosmetic products, including:

- skincare (e.g., moisturizers, lotions, soaps, cleansers);
- haircare (e.g., conditioners, masks, anti-dandruff products);
- make-up (e.g., lipsticks);
- perfumery (e.g., perfumes).

The most important reason for using EOs in cosmetics is their pleasant odor, which effectively masks unpleasant odors associated with fatty acids, fatty oils, and surfactants used in the production of cosmetic products. Such products can be labelled as "fragrance-free", "contains no perfume" or "scented-free" because they do not contain chemical fragrance [13, 19].

Additionally, certain EOs have been used in aromatherapy for various benefits such as mood-enhancing properties. For example, citrus oils like lemon and orange can uplift and energize; chamomile and lavender oils can promote relaxation and stress relief, etc. [13, 19].

6.1.4 Growing demand for EOs in cosmetics

Large quantities of EOs are produced owing to the growing demand of using pure natural ingredients in flavours and fragrances, phytomedicine, aromatherapy, and cosmetics [12, 20, 21]. EOs demand mainly comes from the following markets, food and beverages (35%), fragrances, cosmetics and aromatherapy (29%), household (16%), and pharmaceutical (15%) [12]. The market share of EOs in food and beverages is the largest as compared to other segments because EOs contain health benefits as the natural ingredients present in them impart the essence of their source. Sales data of globally operating flavours and fragrance manufacturers indicate the growing demand for EOs [12]. Orange EO is in high demand in cosmeceutical industries because it offers multiple health benefits like skin firmness, elasticity, and helps in reducing scars, acne, and stretch marks. Lemon and orange EOs are ideal ingredients of skincare and haircare products as they have antiseptic properties.

Many countries do not record domestic production and export data, which makes it difficult to get exact figures for EOs production worldwide. As per the estimate [22], the global EOs production in 2017 was more than 150,000 tonnes valued at about $6B USD [23], which is 50% more since 2007. The major EOs producing countries across the world are China and India, followed by Indonesia, Sri Lanka, and Vietnam. Also, many countries from Africa, the United States, Canada, Mexico, few central and South American countries make sizeable contributions to EOs production while, France, Germany, Japan, Jamaica, and the Philippines also make small contributions.

The top three EOs produced are orange, mint, and lemon. In 2016, about 100,000 tonnes of each was produced, which is more than two-thirds of the total EO crop production. Several EOs are produced on a small scale:

patchouli, litsea, citronella, blue gum (nilgiri), clove leaf (1,000–10,000 tonnes), as well as vetiver, geranium, ylang-ylang, nutmeg, and lavender (50–400 tonnes) [22]. Few EOs are produced on a small scale and sold at higher prices due to less availability [e.g., agarwood oil (6,000–11,000 €/kg), iris (6,200–100,000 €/kg depending on the concentration of irones), or rose oil (6,000–10,000 €/kg)]. These prices vary depending upon the availability of the raw materials, harvesting issues, environmental factors, or yield of extraction [12].

6.1.5 Worldwide trade of EOs

In 2022, the world export of EOs was $5.72B USD and import was $5.72B USD [24]. In 2022, the top exporters of EOs were India ($879M), the United States ($693M), Brazil ($425M), France ($424M), and China ($407M). In 2022, the top importers of EOs were the United States ($1.25B), France ($440M), Germany ($357M), China ($342M), and the Netherlands ($301M).

Though there are benefits of using EOs in cosmetic products, there are some drawbacks associated with them too like, high volatility, poor solubility in water, and their thermal and chemical liabilities, which make their handling difficult for design of cosmetic products. [13, 14]. It is a misconception that use of herbs, herbal extracts, or EOs in formulation makes it safe for consumers. EOs may be toxic as compared to herbs due to their high concentration and/or physicochemical properties [25]. EOs are mixtures of low-molecular weight lipophilic compounds, which enable them to pass across skin membrane easily but may also cause some adverse reactions, e.g. allergies, chronic toxicity in human cells, and photosensitization. Skin structure and its metabolism is influenced by age. Children and elderly persons' skin is delicate and may be vulnerable to toxic effects of the product when applied to skin. Therefore, it is crucial for cosmetic manufacturers to perform the risk assessment of EOs and be aware of the potential implications when including them in cosmetic products [13, 19, 25].

Many industries prefer the economic option of laboratory-developed synthetic fragrances to copy the chemical and aromatic constituents similar to natural oils which are more expensive. However, synthetic fragrances may not deliver the benefits of natural EOs and could be harmful to human application. Synthetic fragrances contain phthalates and benzene derivatives that are known as endocrine disruptors [26] and carcinogens, respectively [12, 27].

As the cosmetics market is dynamic, innovative, and continuously evolving, it must be regulated to ensure safe and quality products are made available to avoid any adverse effects on consumers. To ensure safe and quality products, many countries have established regulatory bodies which are responsible for setting guidelines and standards for safety, efficacy, quality, nomenclature, labelling, packaging, etc., and enforcing them (Figure 6.2). However, regulatory framework differs from country to

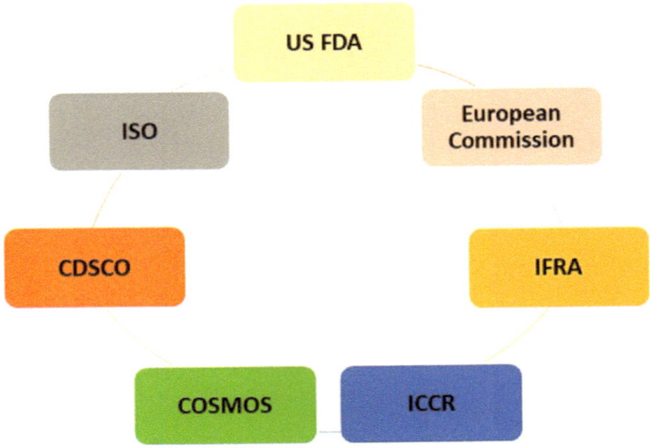

Figure 6.2 Important global regulatory framework.

country, and there is a need to harmonize them. These harmonizing efforts will help the cosmetic industry to sell the same product across all markets [28]. Differences in regulatory framework hamper the safety assessment of cosmetic products and industry growth. Most of the major markets broadly follow similar regulatory norms; however, the existing differences between them restrict growth potential and innovation, thereby negatively impacting the cosmetic industry. Differences in the regulatory framework of each country act as a hindrance to regulatory authorities that ensure every product complies with the regulations used by each country and also affect international trade [29].

Hence, it becomes essential to come up with a solution that will help align the cosmetics regulatory frameworks, support market growth, eliminate trade restrictions, and encourage innovation. To achieve this crucial objective, several international organizations [International Cooperation on Cosmetics Regulation (ICCR), International Organization for Standardization (ISO), Organization for Economic Co-operation and Development (OECD)] have been making concerted efforts in this direction. In 2007, a voluntary group of cosmetic regulatory authorities from the United States, the European Union (EU), Brazil, Japan, the Republic of Korea, Canada, and Chinese Taipei established ICCR. All members of ICCR meet on an annual basis and conduct constructive discussions on various topics related to cosmetic industry trade associations, safety, and regulation (e.g., alternatives to animal testing, nanotechnology, microbiological limits, etc.) [30]. Other organizations like OECD [31] and ISO [32] play a key role in the development of international standards on cosmetics, and mutual acceptance of testing methods guidelines. However, there is still a lot of scope to support the ongoing dialogue and to strengthen the existing cooperation efforts between different countries [33].

As EOs are associated with biological activities, they have been used in many different traditional healing systems around the world, for example, in Ayurvedic medicine, Arabian traditional medicine, traditional Chinese medicine, traditional aboriginal Australian medicines, traditional Korean medicine, and ancient Greek medicine. Herbal medicinal products (HMPs) containing EOs occupy an important place in the modern health care system, especially in the over-the-counter (OTC) i.e. non-prescription segment. EOs are used in HMPs both for human and veterinary use and in traditional herbal medicinal products (THMPs) for human use. The EU has published monographs of EOs as they are used as an active pharmaceutical ingredient (API) to assure the quality, efficacy, and safety of the final medicinal products [13].

European Pharmacopoeia (Ph. Eur.) includes a general Chapter 5.30 on "Monographs on essential oils" (information chapter), a general monograph "Essential oils" (Ph. Eur. 04/2022:2098), and 32 individual monographs on EOs. These monographs describe legally binding quality standards and contain detailed analytical methods for the identification and quality control of EOs. The Committee on Herbal Medicinal Products (HMPC) has developed EU herbal monographs like Ph. Eur., which define requirements for the quality of EOs [34], safety and efficacy reference material, and assessment standards by applicants and National Competent Authorities (NCAs). It also plays a supportive function through well-established use (WEU) marketing authorization and traditional use registration (TUR) [34]. EU list entries are legally binding on applicants and national competent authorities in the Member States with respect to TUR.

6.2 REGULATORY BODIES

6.2.1 European Union

The EU has a well-established and extensive regulatory framework for cosmetics; they are regulated by the Cosmetics Regulation (EC) No. 1223/2009. It is the primary legislation governing cosmetic products in the EU. On 11th July 2013, the Cosmetics Directive (76/768/EEC) was replaced by Cosmetics Regulation (EC) No. 1223/2009 to harmonize and simplify the rules for cosmetics within the EU [35]. This regulation, as amended [36], has certain provisions specifically to address the use of EOs in cosmetic products, most notably in Annex II and Annex III.

6.2.1.1 Key aspects

6.2.1.1.1 Safety

According to the regulation, all cosmetic products placed on the EU market must be safe for human health under normal or reasonably foreseeable conditions of use. The responsibility lies with the manufacturer to ensure the safety of their products and to conduct a safety assessment before placing them on the market.

6.2.1.1.2 Responsible person

According to the regulation, there should be a designated "responsible person" identified within the EU for each cosmetic product. The responsible person must ensure the products they place on the market meet all regulatory compliance including product safety, labelling, claims, and notifications. The responsible person needs to ensure that each cosmetic product has undergone a safety assessment based on the relevant information before placing the cosmetic product on the market. A safety report has to be generated and must be available in accordance with Annex I guidelines, which are laid down in the Commission Implementing Decision 2013/674/EU. The regulation gives clear guidelines about who is the responsible person for cosmetic products that are manufactured within the EU or products that are imported or products that are placed on the market with distributors' trademarks [37]. Regulation contains clear guidelines on the obligations of the responsible person (a) if they have a reason to believe that a cosmetic product which they have placed on the market is not in conformity with this Regulation, responsible person shall immediately take the corrective measures necessary to bring that product into conformity, withdraw it or recall it, as appropriate; (b) if the cosmetic product presents a risk to human health, responsible persons shall immediately inform the competent national authorities of the Member States for further necessary action.

6.2.1.1.3 Product information file (PIF)

The responsible person must maintain a product information file (PIF) for each cosmetic product. The PIF should include details about product formulation, compliance statement with GMP, safety assessment, method of manufacturing, proof of claims, and data on animal testing. PIF of each cosmetic product shall be kept for a period of ten years following the date on which the last batch of the cosmetic product was placed on the market.

6.2.1.1.4 Notification

Before placing a cosmetic product on the EU market, the responsible person must notify the European Commission through the cosmetic products notification portal. The notification should include information about the product, its name, specific identification, country of origin if the product is imported, the responsible person and his contact, name of Member State, the manufacturer, and the product's safety. Once the product is placed on the market, the responsible person shall notify the Commission of the original labelling and a photograph of the corresponding packaging.

6.2.1.1.5 Labelling

Labelling is an important aspect of any cosmetic product and should be labelled in compliance with the regulations. The label should include information such as the name and address of the responsible person, the nominal content at the time of packaging, quantity, the period after opening (PAO), batch number, function of the product, a list of ingredients (in descending order of weight of the ingredients), particular precautions to be observed.

6.2.1.1.6 Ingredient restrictions

Detailed guidelines are outlined for the use of specific ingredients in cosmetic products. Annexes II, III, IV, V, and VI of the regulation provide lists of prohibited substances, restricted substances, colourants, preservatives, and ultraviolet (UV) filters, respectively. Annex II to Regulation (EC) No. 1223/2009, as amended, lists substances that are prohibited for use in cosmetic products. Among substances listed in this annex are:

- Plant ingredients which are prohibited in cosmetic products, regardless of the function.
- Plants and their compounds that are prohibited in cosmetic products for a given function (perfume ingredients)
- Substances that are prohibited in cosmetic products unless they are naturally present in extracts and EOs which are subject to concentration limits.

The Scientific Committee on Consumer Safety (SCCS) published an opinion on fragrance allergens in cosmetic products in June 2012 [6 European Commission, SCCS opinion on fragrance allergens in cosmetic products, 2012 (SCCS/1459/11).]. In this opinion, EOs and their effects as contact allergens in humans are evaluated. The latest study on fragrance allergen labelling on cosmetic products that was commissioned by the European Commission was published in November 2020 [38].

Specific national legislation and recommendations are as below:

France

- Recommendations regarding EOs quality criteria (May 2008) [39].
- Recommendations for the risk assessment associated with the use of EOs in cosmetic products (October 2010) [40].

Germany

- The Federal Institute for Risk Assessment (BfR) recommended the following [41] maximum concentration of 1% of camphor, eucalyptus oil, menthol, or methyl salicylate in leave-on products and a maximum

concentration of 5% of camphor, 4% of menthol, and 2.5% of methyl salicylate in rinse-off products [42].
- The BfR has published an opinion on tea tree oil recommending a concentration limit of 1% for tea tree oil in cosmetic products [43].

6.2.1.2 Recommendations of the Council of Europe

As there is an increase in the number of plant-based products in the market, the Council of Europe has published three volumes of recommendations concerning the use of plants and plant-based preparations as ingredients in cosmetic products [44–46]. It has published a safety survey of ingredients used in cosmetic products [47] and has recommended avoiding the use of camphor, eucalyptol, or menthol in cosmetic products for infants [48].

The Council of Europe's Committee of Experts on Cosmetic Products has evaluated several datasheets on plant and plant preparations and are included in these volumes. The entries are further classified into three categories: (a) plants without health hazards; (b) plants for which further information is needed; and (c) plants that may pose a health risk and are not recommended for use in cosmetic products.

The Committee highlights some of the raw materials contain substances with significant activity, which could be potentially harmful to consumers. Hence, it is necessary to do a risk assessment of these substances while evaluating the overall safety of a cosmetic product.

6.2.1.3 Recommendations of the International Fragrance Association (IFRA)

The International Fragrance Association (IFRA) Scientific Committee collects data and makes it available for the safety evaluation of fragrance ingredients. IFRA's information should be taken into account by operators incorporating EOs into cosmetic products because many fragrances contain EOs and their components.

6.2.1.3.1 Claims

The European Commission has published guidelines (EC) No. 1223/2009, Article 20 on cosmetic product claims to help compliance. The regulation has set guidelines for claims made on cosmetic products to ensure that they are truthful, clear, and substantiated. In addition, the regulation indicates Commission should define the "Action Plan" to standardize the declarations applied to cosmetics to set common criteria for its use [22].

6.2.1.3.2 Good manufacturing practices (GMP)

The European Committee for Standardization has published the standard EN ISO 22716:2007, which provides guidance on good manufacturing practices (GMP) for cosmetics. As per regulation, cosmetic products should be manufactured following GMP guidelines to ensure product quality and safety.

6.2.1.3.3 Nanoparticles

The regulation has specific provisions ((EC) No. 1223/2009, Article 16) for the use of nanomaterials in cosmetic products and recommends taking high level of protection of human health [37]. All ingredients present in the form of nanomaterials shall be clearly indicated in the list of ingredients. The names of such ingredients shall be followed by the word "nano" in brackets. Cosmetic products containing nanomaterials shall be notified to the Commission by the responsible person by electronic means six months before being placed on the market. Up to now, the Commission has authorized four UV filters as nanomaterials: methylene bis-benzotriazolyl tetramethylbutylphenol (MBBT), titanium dioxide, zinc oxide, and tris-biphenyl triazine. It has also allowed carbon black (nano) for use as a colourant in cosmetic products [49].

There was a corrigendum to Commission Regulation (EU) 2024/858 of 14th March 2024 amending Regulation (EC) No. 1223/2009 of the European Parliament and of the Council as regards the use of the nanomaterials styrene/acrylates copolymer, sodium styrene/acrylates copolymer, copper, colloidal copper, hydroxyapatite, gold, colloidal gold, gold thioethylamino hyaluronic acid, acetyl heptapeptide-9 colloidal gold, platinum, colloidal platinum, acetyl tetrapeptide-17 colloidal platinum and colloidal silver in cosmetics products [50]. From 1st February 2025, cosmetic products containing these substances and not complying with the restrictions shall not be placed on the Union market. From 1st November 2025, cosmetic products containing these substances shall not be made available on the Union market.

6.2.1.3.4 Animal testing

The EU regulation Article 18 prohibits the testing of cosmetic products and their ingredients on animals for both safety and efficacy assessments [51].

6.2.2 United States

In the United States, cosmetics are regulated under two acts, the Federal Food, Drug, and Cosmetic Act (FD&C Act) and the Fair Packaging and Labelling Act (FPLA) [52, 53]. The US Food and Drug Administration (FDA) regulates cosmetics under the authority of these laws and is responsible for overseeing and enforcing these laws. However, the FDA does not pre-approve cosmetic products before they are marketed, except for colour additives. Herbal cosmetics are also covered under the same regulations. United States federal law incorporated FPLA in 1966 to protect consumers and provide standardized information on product packaging and labels with the aim of facilitating value comparisons and preventing unfair or deceptive packaging and labelling of consumer commodities. The US FDA and the Federal Trade Commission (FTC) are responsible for enforcing the FPLA.

6.2.2.1 Key aspects

6.2.2.1.1 Identity of the product

The name of the product should clearly and prominently be displayed on the label or a brief description of its contents or nature [54]. The following information must appear on the principal display panel: (a) an identity statement, indicating the nature and use of the product, (b) an accurate statement of the net quantity of contents. In addition to this, the name and place of business, directions for safe use, warning and caution statements, ingredients, and if the product is also a drug, its labelling must comply with the regulations for both OTC drug and cosmetic ingredient labelling [55]. The label on each package of a cosmetic should include the name of each ingredient in descending order of predominance, except that fragrance or flavour may be listed as fragrance or flavour [56].

6.2.2.1.2 Name and place of business

The label must include the name and place of business of the manufacturer, packer, or distributor. If the name and address are not those of the manufacturer, the label must say "Manufactured for..." or "Distributed by...", or similar wording expressing the facts [21 CFR 701.12(c)].

6.2.2.1.3 Net quantity of contents

The label must show the net quantity of the product in terms of weight, measure, or numerical count. The quantity should be expressed in both ounces and pounds [57]. The net quantity should be displayed prominently and in a way that is easy for consumers to read.

6.2.2.1.4 Ingredients

The US FDA gives exhaustive information on cosmetics labelling and ingredients. Cosmetic products must include a list of ingredients in descending order of predominance [52]. Ingredients, other than colour additives, which are present at a concentration greater than 1%, should be in descending order of predominance, followed by ingredients, other than colour additives, present at a concentration of not more than 1%.

6.2.2.1.5 Product safety

Each ingredient used in a cosmetic product and each finished cosmetic product shall be adequately substantiated for safety before marketing. Cosmetic manufacturers are responsible for ensuring the safety of their products. The FDA does not pre-approve cosmetic products or their ingredients, except for colour additives. Manufacturers must give evidence for the safety of their products and ingredients used for manufacturing before marketing them [58].

6.2.2.1.6 Prohibited and restricted ingredients

The FDA does not allow the use of prohibited ingredients in cosmetics or any ingredient that makes the product harmful when consumers use it according to directions on the label, except for coal-tar hair dyes, which the law treats differently. Some ingredients are prohibited, while others are allowed only under specific conditions or in limited amounts. A list of prohibited and restricted ingredients can be found in the Code of Federal Regulations Title 21, parts 700–740 [52].

6.2.2.1.7 Colour additives

The FDA regulates the use of colour additives in cosmetics, and pre-market approval is required for these additives. FDA has given an exhaustive list of colour additives that are exempt from certification and permitted for use in cosmetics (21 CFR Part 73 Subpart C—Cosmetics) and colour additives that are subject to certification and permitted for use in cosmetics (21 CFR Part 74 Subpart C—Cosmetics and 21 CFR Part 82 Subparts B, C, and D). Colour additive violations are a common reason for detaining imported cosmetic products offered for entry into this country [59].

6.2.2.1.8 Labelling

The FDA regulates cosmetic labelling under the authority of both the FD&C Act and the FPLA. The label should include information such as the identity of the product, the name and place of business of the manufacturer, packer, or distributor, and the net quantity of contents before product is out on the market. Labelling must be truthful and not misleading. The FDA also requires proper labelling of ingredients in descending order of predominance [55, 60].

6.2.2.1.9 GMP

Rigorous adherence to GMP minimizes the risk of adulteration or misbranding of cosmetics. The FDA provides guidance to industry and other stakeholders (e.g., consumer interest groups, academia, and other regulatory groups) on FDA's current thinking concerning what constitutes GMP for cosmetics.

These guidelines are intended to assist the industry and other stakeholders in identifying the standards and issues that can affect the quality of cosmetic products [61]. The FDA has revised the "Cosmetic Good Manufacturing (GMP) Guidelines/Inspection Checklist" by updating the current practice and clarifying certain topic areas based on recent experiences. As part of an international harmonization effort with ICCR, the FDA has agreed to consider the current ISO standard for cosmetic GMP (ISO 22716:2007) when revising this guidance. The FDA reviewed ISO 22716 and decided to incorporate, modify, or exclude specific aspects of it, into/from this guidance based on its experience.

Although the FDA does not have specific GMP regulations for cosmetics, manufacturers are expected to adhere to GMP guidelines to ensure product safety and quality. The FDA provides guidance on cosmetic GMP through the publication of "Cosmetic GMP guidelines" [62].

6.2.2.1.10 Adverse event reporting

Cosmetic manufacturers and distributors are encouraged to report adverse events related to their products voluntarily. The FDA's Center for Food Safety and Applied Nutrition Adverse Event Reporting System is used to collect and monitor such reports. These reports are very helpful to the agency when identifying safety signals and choosing particular products for further scrutiny [63].

6.2.2.1.11 Claims

As per the FDA, information on cosmetic labelling, including claims, must be truthful and not misleading. Also, the product should not be marketed with claims for purposes such as treating or preventing disease or affecting the structure or function of the body including the skin because then the product may be considered as a drug and subject to additional regulations. Cosmetic products must not make claims that would classify them as drugs under the FD&C Act [64].

6.2.3 International Organization for Standardization (ISO)

ISO is an independent, non-governmental international organization that develops and publishes international standards covering various industries, including cosmetics. ISO has several standards related to cosmetics such as ISO 22716 for GMP and ISO 16128 for natural and organic cosmetic ingredients [65]. The guidelines related to GMP have been prepared considering the specific needs of the cosmetic sector. These guidelines offer organizational and practical advice on the management of the human, technical, and administrative factors affecting product quality. ISO Technical

Committee for Standardization in the field of cosmetic products (ISO/TC 217) has contributed 27 standards to the sustainable goals out of a total of 49 published ISO standards and 3 ISO standards currently under development [66].

ISO 16128-2:2017 describes guidelines on definitions and criteria for natural and organic cosmetic ingredients and products. These guidelines are specific to the cosmetics sector, which apply scientific judgement and offer principles towards a consistent logical framework for natural and organic cosmetic ingredients and products incorporating common approaches [67].

ISO codifies the composition of each EO with an individual identification [e.g., ISO 1342:2012 for *Rosmarinus officinalis* L. (rosemary), ISO 3061:2008 for *Piper nigrum* L. (black pepper) and ISO 3518:2022 for *Santalum album* L. (sandalwood)]. In ISO/Technical Committee 54, ISO has standardized several analytical methods for quality control of these specifications along with requirements for labelling, transport, and marking. ISO/TC 54 emphasizes global trade in EOs, thus focusing on improving the quality of the EO market, protecting the health of consumers who buy products containing EOs, and increasing the safety level of these oils [12, 68]. EOs need to be protected from light and the container should be tightly closed and fully filled; otherwise, it may undergo chemical reactions like photo isomerization, photo cyclization, oxidation, peroxidation, and decomposition of alcohols, and ketone hydrolysis. The safety of EOs or single compounds may alter due to improper storage conditions. The ISO/TS 210:2014 standard prescribes general rules for the packaging, conditioning, and storage of EOs [19].

Twenty-six possible allergenic fragrances have been defined, out of which 18 can be found as ingredients of EOs [69]. Due to this, they must be declared on the packaging or in the information leaflet if the concentration of these allergenic fragrances is higher than the permissible concentration of 0.01% in shower gels and baths (rinse-off products) and higher than 0.001% in body oils, massage oils, and creams (leave-on products).

6.2.4 International Fragrance Association (IFRA)

The International Fragrance Association (IFRA) represents the interests of the fragrance industry worldwide [70]. Membership is open to associations of fragrance manufacturers from all countries/regions and currently includes members from Asia/Pacific, Europe, and the Americas. All IFRA members have committed to comply with the IFRA Code of Practice, which aims to provide recommendations for (a) ensuring safe use and regulatory compliance, (b) applying the IFRA Standards, and (c) adhering to good operating practices, and includes fragrance safety standards that may limit or ban the usage of certain fragrance materials [71]. IFRA clearly mentions that it is the responsibility of each IFRA member to ensure that the fragrance ingredients and mixtures they supply comply with applicable laws and are safe for their intended uses. IFRA Standards are prepared by its Risk Management Task

Force on the basis of safety assessments prepared by the Research Institute for Fragrance Materials (RIFM) and decisions made by an independent Expert Panel for Fragrance Safety. IFRA also notifies new and amended Standards to IFRA members, associations, and other stakeholders.

The IFRA defines which EOs and which components represent a potential allergy risk and determines their maximum concentration to produce safe cosmetic products [19]. IFRA also issues recommendations for the safe use of fragrance ingredients, which are published in the IFRA Code of Practice and its guidelines [72].

6.2.5 Association of South East Asian Nations (ASEAN)

The ASEAN Cosmetics Association, Inc. (ACA) is an organization of different countries that aims to harmonize and maintain cosmetics regulations, upgrade the standards of technical competence and scientific expertise in the local industries, and expand the cosmetics trade in ASEAN. The ASEAN has harmonized its cosmetic regulations through the ASEAN Cosmetic Directive (ACD), based on the EU cosmetics regulation (EC) No. 1223/2009 [73]. The ACD covers requirements such as product safety, labelling, ingredient restrictions, claims, and analysis. Herbal cosmetics are also subject to the ACD [74]. The aim of the ACD is to eliminate trade barriers, promote the free movement of goods, and ensure product safety and quality within the member countries.

6.2.5.1 Key aspects of the ASEAN Cosmetic Directive

6.2.5.1.1 Product safety

Cosmetic products that are marketed in the ASEAN region shall be safe for human health under normal or reasonably foreseeable conditions of use. Manufacturers are responsible for product safety and conducting its safety assessments before placing the product on the market.

6.2.5.1.2 Cosmetic product notification

The manufacturer or importer must notify the relevant national authority before a cosmetic product can be marketed in the ASEAN region. The notification should also include information about the product, the manufacturer or importer, and the safety of the product.

6.2.5.1.3 Labelling

As per ACD, cosmetic product labels must be in full compliance with the ACD. The label should include information such as the trademark, name, and address of the manufacturer or importer, country of manufacture, nominal content, the PAO, conditions for use, batch number, function of the product, and a list of ingredients.

6.2.5.1.4 Ingredient restrictions

The ACD guidelines for the use of specific ingredients in cosmetic products are based on the EU cosmetics regulations. According to them, Annexes II, III, IV, V, and VI are lists of prohibited substances, restricted substances, colourants, preservatives, and UV filters, respectively [75].

6.2.5.1.5 Claims

The ACD gives guidelines for claims made on cosmetic products and instructs the Member States to ensure that they are truthful, clear, and have justified evidence.

6.2.5.1.6 Herbal cosmetics in ASEAN

Herbal cosmetics are bound by the same regulations as conventional cosmetics under the ACD. Herbal ingredients used in cosmetic products in the ASEAN region must abide by the restrictions and prohibitions given by the ACD and must be evaluated for safety. Additionally, the national regulatory body of each country may also provide guidelines on the use of herbal ingredients depending on their specific regulations and traditional medicine systems.

6.2.6 India

In India, cosmetics products are regulated under two acts, the Drugs and Cosmetics Act of 1940, and the Drugs and Cosmetics Rules of 1945 [76]. The regulations in India cover guidelines for cosmetic products, including manufacturing, labelling, ingredient restrictions, and safety. In India, herbal cosmetics are also subject to the same regulations; additionally, guidelines are provided by the Indian Pharmacopoeia Commission under the Ayurvedic, Siddha, and Unani (ASU) systems [77, 78]. The Central Drugs Standard Control Organization (CDSCO) is the primary regulatory body overseeing the safety, quality, and efficacy of cosmetics in India [79].

6.2.6.1 Key aspects of Indian cosmetic and herbal cosmetic regulations

6.2.6.1.1 Product safety

According to CDSCO, cosmetic products marketed in India must be safe for human use and abide by the provisions of the Drugs and Cosmetics Act, 1940, and the Drugs and Cosmetics Rules, 1945. Cosmetic manufacturers are responsible for ensuring the safety and quality of products. The latest amendments to Cosmetic Rule 2020 are applied to the regulation of cosmetics in India [79].

6.2.6.1.2 Licensing

Cosmetic manufacturers in India are required to obtain a manufacturing license from the state licensing authority. The manufacturers must comply with the Drug and Cosmetic Rules, 1945, Schedule M-II, which give specific requirements regarding manufacturing premises and set-up.

6.2.6.1.3 Labelling

All cosmetic product labels must be in compliance with the Drugs and Cosmetics Rules, 1945. Labels should include necessary information such as the product's name, manufacturer's name and address, manufacturing license number, batch number, manufacturing date, expiration date, and a list of ingredients.

6.2.6.1.4 Ingredient restrictions

The Drugs and Cosmetics Rules, 1945, gives information regarding the use of specific ingredients in cosmetic products. Specifically, appendixes A and B outline the standards for cosmetics and the list of prohibited substances, respectively.

6.2.6.1.5 Import and registration

Before importing cosmetics in India, the importer of cosmetics must obtain a registration certificate from CDSCO. The registration process is well explained, and the importer should submit relevant documents, including product information, safety data, and details of the manufacturing site.

6.2.6.1.6 Herbal cosmetics in India

Herbal cosmetics in India are regulated by the same regulations as conventional cosmetics under the Drugs and Cosmetics Act, 1940, and the Drugs and Cosmetics Rules, 1945. The herbal cosmetics in India must comply with the regulations regarding the safety, quality, and efficacy of herbal ingredients used in cosmetics. In addition, the Ministry of Ayurveda, Yoga, Naturopathy, Unani, Siddha, and Homoeopathy provides guidelines and standards for ASU drugs, which may be applicable to herbal cosmetics containing ingredients derived from these traditional medicine systems.

6.2.7 China

In China, cosmetic products are regulated and governed by the National Medical Products Administration (NMPA) [80]. The NMPA works under the jurisdiction of the State Administration for Market Regulation (SAMR), and it is the primary regulatory body responsible for the supervision, management, and administration of drugs, medical devices, and cosmetics in China. The regulatory framework gives clear guidelines for cosmetic product registration, approval, monitoring, safety assessment, labelling, and GMP. Herbal cosmetics are also subject to the same regulations, with additional guidelines for the use of herbal ingredients [81, 82].

6.2.7.1 *Key aspects of the NMPA's regulation of cosmetics in China*

6.2.7.1.1 Cosmetic classification

The NMPA classifies cosmetics into two categories: nonspecial use cosmetics (NSUC, e.g., skincare, haircare, and makeup products) and special use cosmetics (SUC, e.g., hair growth, hair dye, hair removal, fitness, deodorizing, spot removal, sun protection, and whitening products). According to the NMPA, nonspecial cosmetics require notification, while special cosmetics require registration [81]. Recently, the NMPA implemented a new Cosmetic Supervision and Administration Regulation (CSAR) in January 2021 to streamline these procedures and improve the regulatory framework [83]. The NMPA maintains a list of prohibited and restricted ingredients for use in cosmetic products while inventory of existing cosmetic ingredients in China (IECIC) contains approved ingredients that can be used in cosmetic formulations. According to the NMPA, cosmetic product labelling should comply with the regulation including information about the product, manufacturer, ingredients, date of manufacturing and expiration, and any necessary precautions or warnings. Manufacturers must comply with the regulatory guidelines including product registration, production license, safety assessment, and GMP to ensure product quality and safety for human use [80].

6.2.8 Japan

In Japan, the Ministry of Health, Labor, and Welfare regulates cosmetics. The Pharmaceutical Affairs Law (Law No. 145 of 1960), and its related regulations provide clear requirements of labelling, safety standards, and ingredient restrictions for cosmetic products in Japan [84].

6.2.9 South Korea

The Ministry of Food and Drug Safety regulates cosmetics in South Korea. The Cosmetic Act Regulations provide guidelines on cosmetic registration, manufacturing, distribution, safety, and labelling rules [85].

6.2.10 Australia

Australian Industrial Chemicals Introduction Scheme (AICIS) is responsible for the regulation of cosmetics in Australia. The Industrial Chemicals (Notification and Assessment) Act 1989 (ICNA Act) and the Australian consumer law both set the regulatory framework for cosmetic products, including labelling, ingredients, and safety standards [86, 87]. AICIS also regulates organic and natural ingredients [88].

6.2.11 Canada

Cosmetics in Canada are regulated by Health Canada regulations under the Food and Drugs Act. Cosmetic regulations set requirements for cosmetic product safety, ingredient restrictions, GMP, packing, and labelling [89].

6.2.12 Brazil

In Brazil, the Brazilian Health Regulatory Agency is responsible for regulating cosmetics. The Collegiate Board Resolution No. 7/2015 and other regulations define the guidelines for cosmetic products, including safety, manufacturing, labelling, and ingredient restrictions [90, 91].

6.2.13 Russia

In Russia, cosmetics are regulated by the Federal Service for Surveillance in Healthcare (Roszdravnadzor). The requirements for cosmetic products, including safety standards, ingredient restrictions, and labelling are set by the Eurasian Economic Union (EAEU) technical regulation on the safety of perfumery and cosmetic products (TR EAEU 009/2011) [92, 93].

There are several other regulatory authorities and guidelines around the world that govern the safety, quality, and labelling of cosmetic products. Some of these include:

6.2.14 International Cooperation on Cosmetics Regulation (ICCR)

ICCR is a voluntary international group of cosmetics regulatory authorities from Brazil (Brazilian Health Regulatory Agency, ANVISA) Canada, Chinese Taipei (Taiwan Food and Drug Administration, TFDA), the European Union, Israel (Ministry of Health of Israel, MOH), Japan, the Republic of Korea (Ministry of Food and Drug Safety, MFDS) and the United States [94]. Each member holds the responsibility for the implementation of ICCR initiatives, and also it successfully conducts constructive dialogue with the regulated community, represented by the cosmetics industry trade associations and other stakeholders.

ICCR's mission is to maintain the highest level of global consumer protection and minimize regulatory barriers to international trade. It engages in information sharing, regulatory cooperation, training, cosmetic safety, and other activities aimed at advancing science and knowledge internationally. Recently 18th Annual Meeting (ICCR-18) was held from 9th to 11th July 2024 in Taipei City. In 2024, representatives from Egypt, Saudi Arabia, Thailand, and the United Kingdom participated in the meeting as observers [95].

6.2.15 Cosmetic Organic and Natural Standard (COSMOS)

The Cosmetic Organic and Natural Standard (COSMOS) is a non-profit international standard organization for organic and natural cosmetics. The COSMOS standard provides guidelines for cosmetic companies, and promotes organic agriculture, to ensure that their products are made using natural resources responsibly and following sustainable practices so that the products are safe for consumers. The COSMOS standard defines what "natural" and "organic" mean in the context of cosmetic products. COSMOS ORGANIC contains the required percentage of organic ingredients as specified in the COSMOS standard, while COSMOS NATURAL complies with the COSMOS standard in all respects but does not meet the required minimum organic percentage as specified. The COSMOS standard covers various aspects in detail like raw material, sourcing, manufacturing, and marketing [96].

6.2.16 Personal Care Products Council (PCPC)

The Personal Care Products Council (PCPC) is a US-based trade association. It represents cosmetic and personal care product manufacturers and provides guidance to ensure product safety, quality, and compliance with regulatory requirements [97].

6.2.17 European Cosmetic Industry Association (Cosmetics Europe)

Cosmetics Europe is a trade association of the cosmetic and personal care industry in Europe, which provides guidance on regulatory compliance, industry trends, and best practices [98].

6.2.18 International Federation of Societies of Cosmetic Chemists (IFSCC)

The International Federation of Societies of Cosmetic Chemists (IFSCC) is a global organization dedicated to international cooperation in cosmetic science and technology, which shares scientific knowledge, research, and best practices in the cosmetic industry [99].

6.2.19 Society of Cosmetic Scientists (SCS)

The Society of Cosmetic Scientists (SCS) is a professional organization dedicated to promoting education, research, and collaboration in cosmetic science. It is affiliated with IFSCC [100] and plays a crucial role in setting standards, promoting best practices, and providing resources to ensure the safety, quality, and efficacy of cosmetic and herbal cosmetic products worldwide.

6.2.20 International Nomenclature of Cosmetic Ingredients (INCI)

The International Nomenclature of Cosmetic Ingredients (INCI) is a standardized naming system for cosmetic ingredients used globally. It was developed by the International Nomenclature Committee (INC) and published by the PCPC to ensure clear communication and transparency about the ingredients used in cosmetic products [101].

6.2.21 The European Federation for Cosmetic Ingredients (EFfCI)

The European Federation for Cosmetic Ingredients (EFfCI) is a European trade association that brings together European manufacturers of synthetic and natural ingredients for the cosmetics and personal care industry. It provides information on international regulations and legal, technical, and scientific developments [102].

6.2.22 The Natural and Organic Cosmetics Association (NATRUE)

The Natural and Organic Cosmetics Association (NATRUE) is a Brussels-based international non-profit organization that establishes criteria for natural and organic cosmetic products. The NATRUE Label is a reliable, verifiable benchmark that provides a certification system so that consumers identify authentic natural and organic cosmetic products [103].

6.2.23 The Vegan Society

The Vegan Society is an international organization that promotes vegan products and provides certification, including cosmetic products. The Vegan Trademark certification ensures that a product is free from animal-derived ingredients and has not been tested on animals [104].

6.2.24 The Coalition for Consumer Information on Cosmetics (CCIC)

The Coalition for Consumer Information on Cosmetics (CCIC) provides a leaping bunny logo, which offers a certification system for cruelty-free cosmetic products by avoiding animal testing in all stages of product development [105].

There are a few more organizations and associations related to the cosmetic and personal care industry.

6.2.25 Ecocert

Ecocert is an international certification body that specializes in organic and natural products, including cosmetics. It has strict ecological and environmental standards [106].

6.2.26 Soil Association

Soil Association is a UK-based organization that promotes organic farming and sustainable practices and offers a certification programme for organic and natural cosmetic products [107].

In addition to this, there are other organizations that support and provide guidance to the cosmetic industry on ingredients, manufacturing, GMP, safety, labelling, research, sustainable practices, quality, and regulatory requirements. Table 6.1 provides the names of global and regional organizations of cosmetic and personal care industry.

Table 6.1 Global and regional organizations of cosmetic and personal care industry

Sr. no.	*Name of organization*	*Ref.*
1	Cosmetic Valley	[108]
2	Cosmetics Alliance Canada	[109]
3	Independent Beauty Association (IBA)	[110]
4	The Indian Beauty and Hygiene Association (IBHA)	[111]
5	Bundesverband der industrie- und handelsunternehmen (BDIH)	[112]
6	The Organic Materials Review Institute (OMRI)	[113]
7	The Cosmetic Ingredient Review (CIR)	[114]
8	The Green Chemistry and Commerce Council (GC3)	[115]
9	The Global Cosmetics Cluster (GCC)	[116]
10	The CTPA	[117]
11	The Australian Society of Cosmetic Chemists (ASCC)	[118]
12	The Society of Cosmetic Chemists (SCC)	[119]
13	The Personal Care Product Council's Science Symposium	[120]
14	The Professional Beauty Association (PBA)	[121]
15	The International Association for Soaps, Detergents, and Maintenance Products (AISE)	[122]

6.3 CONCLUSION

In this era, the cosmetic industry needs to release innovative products regularly using more natural ingredients like EOs due to the multiple benefits associated with them. Although EOs are used in diluted form in cosmetics, they may still have adverse effects on sensitive customers. It is therefore important that cosmetic innovators and manufacturers try to establish quality parameters for using EOs before placing the final product on the market.

Brands are more focused on the use of scientific innovation and technology for competitive advantage. Successful products in the market make use of the latest innovations, and biotechnology in ingredient design, product development, packaging, and storytelling. This chapter provides a quick overview of the important global regulatory framework that governs cosmetics and herbal cosmetic products. By following relevant applicable regulations, the cosmetic industry ensures product quality and safety and protects consumer interest. Also, the cosmetic industry must keep itself well informed regarding updates on the changing regulatory landscape and its harmonizing efforts.

REFERENCES

1. Mishra, L., Kurmi, B. D. Cosmetics Regulations and Standardization Guidelines. *Pharmaspire*, 2023, *15*, 137–150.
2. https://cdsco.gov.in/opencms/opencms/en/Cosmetics/cosmetics/#:~:text=Cosmetic%20is%20defined%20under%20section,or%20altering%20the%20appearance%2C%20and] [accessed on 03-Sept-2024].
3. Precedence Research. Cosmetics Market, 2024(By category: skin & sun care products, hair care products, deodorants & fragrances, makeup & colour cosmetics; by gender: men, women, unisex; by distribution channel: hypermarkets/supermarkets, specialty stores, pharmacies, online sales channels, other) – Global Industry Analysis, Size, Share, Growth, Trends, Regional Outlook, and Forecast 2023–2032. Available online https://www.precedenceresearch.com/cosmetics-market] [Last accessed on 03-Sept-2024].
4. Bundesinstitut für die Risikobewertung (BfR), FAQs, 'Frequently Asked Questions about the Use of Essential Oils', 28.02.2008 [Last accessed on 16-Oct-2024].
5. European Pharmacopoeia: Essential oils, 2016. https://www.edqm.eu/en/guidance-on-essential-oils-in-cosmetic-products [Last accessed on 16-Oct-2024].
6. ISO 9235: 2013. Aromatic Natural Raw Materials – Vocabulary. https://www.iso.org/obp/ui#iso:std:iso:9235:dis:ed-3:v1:en
7. Sharma, A., Kumar, V., Mittal, C., Rana, V., Dabral, K., Parveen, G. Role of Essential Oil Used Pharmaceutical Cosmetic Product. *J. Res. Appl. Sci. Biotech.* 2023, 2, 147–157.
8. Butola, K., Bisht, V., Kumar, R. Recent Approaches of Ocular Disease and Its Herbal Product Treatment: An Updates. *J. Res. Appl. Sci. Biotech.* 2023, 2, 102–114.
9. Dongare, P. N., Bakal, R. L., Ajmire, P. V., Patinge, P.A., More, M. P., Manwar, J. V. An Overview on Herbal Cosmetics and Cosmeceuticals. *Int. J. Pharm. Sci. Rev. Res.* 2021, *68*, 75–78.

10. Ratajczak, P., Landowska, W., Kopciuch, D., Paczkowska, A., Zaprutko, T., Kus, K. The Growing Market for Natural Cosmetics in Poland: Consumer Preferences and Industry Trends. *Clin. Cosmet. Investig. Dermatol.* 2023, *16*, 1877–1892.
11. Hussain, H., Al-Harrasi, A., Green, I. R. Chapter 48. Frankincense (Boswellia) Oils. In Essential Oils in Food Preservation, Flavour and Safety, Preedy, V. R., Ed., Elsevier: Amsterdam, The Netherlands, 2016, pp. 431–440.
12. Sharmeen, J. B., Mahomoodally, F. M., Zengin, G., Maggi, F. Essential Oils as Natural Sources of Fragrance Compounds for Cosmetics and Cosmeceuticals. *Molecules.* 2021, *26*, 666.
13. Petina, Olga. "Comparative analysis of the EU regulatory requirements for essential oils as ingredients for the pharmaceutical, cosmetic products and feed additives." Master's thesis for "Master of Drug Regulatory Affairs, M.D.R.A." The Faculty of Mathematics and Natural Sciences, The Rheinische Friedrich-Wilhelms-Universität Bonn, Hamburg, 2023, and references cited therein. https://www.dgra.de/media/pdf/studium/masterthesis/master_petina_olga_2023.pdf
14. Guzmán, E., Lucia, A. Essential Oils and Their Individual Components in cosmetic Products. *Cosmetics.* 2021, *8*, 114.
15. Babita, S., Sellam, P., Jayoti, M., Puja, R. Floral Essential Oils: Importance and Uses for Mankind. *HortFlora Res. Spectr.* 2014, *3*, 7–13.
16. Franz, C., Novak, J. Sources of Essential Oils. In Handbook of Essential Oils and Science, Technology and Applications, Hüsnü Can Baser, K., Buchbauer, G., Eds., CRC Press Taylor & Francis Group: Boca Raton, FL, 2010, pp. 39–81.
17. Graf, M., Stappen, I. Beyond the Bark: An Overview of the Chemistry and Biological Activities of Selected Bark Essential Oils. *Molecules.* 2022, *27*, 7295.
18. Bolouri, P., Salami, R., Kouhi, S., Kordi, M., Lajayer, B. A., Hadian J, Astatkie T. Applications of Essential Oils and Plant Extracts in Different Industries. *Molecules.* 2022, *27*, 8999.
19. Sarkic, A., Stappen, I. Essential Oils and Their Single Compounds in Cosmetics—A Critical Review. *Cosmetics*, 2018, *5*, 11.
20. Baser, K. H. C., Buchbauer, G. Handbook of Essential Oils: Science, Technology, and Applications, CRC Press: Boca Raton, FL, 2015, 35.
21. Baser, K. H. C., Demirci, F. Chemistry of Essential Oils. In Flavours and Fragrances: Chemistry, Bioprocessing and Sustainability, Berger, R. G., Ed., Springer: New York, NY, 2007, pp. 43–86.
22. Barbieri, C., Borsotto, P. Essential Oils: Market and Legislation. In Potential of Essential Oils, El-Shemy, H., Ed., IntechOpen: London, 2018, pp. 107–127.
23. What future for essential oils. Chinaeasa Zhuhai, May 2017. In Definition y Caracterization delos Aceitesesenciales Naturales [Definition and characterization of Natuiral Essential Oils]. Agencia Espanola de Medicamentos y Productos Sanitarios (AEMPS). Madrid 29 Junio 2017.
24. https://oec.world/en/profile/hs/essential-oils. [Last accessed on 20-Sept-2024].
25. European Directorate for the Quality of Medicines & HealthCare. Guidance on Essential oils in Cosmetic Products. EDQM, Strasbourg, 2016. https://www.edqm.eu/en/guidance-on-essential-oils-in-cosmetic-products.
26. Olujimi, O., Fatoki, O. S., Odendaal, J. P., Okonkwo, J. Endocrine disrupting chemicals (phenol and phthalates) in the South African environment: A need for more monitoring. *Water SA* 2010, *36*, 5.
27. Carson, P. A. Hazardous Chemicals Handbook, Elsevier: Amsterdam, The Netherlands, 2002.
28. Renner, G., Jacques, M. An Overview of Existing Cosmetics Frameworks: Common Principles and Compatibility. *Cosmet. Eur.* 2021, *16*, 38–40.

29. Risk & Policy Analysts Limited. Comparative Study on Cosmetics Legislation in the EU and Other Principal Markets with Special Attention to So-Called Borderline Products, Risk & Policy Analysts Limited: Norflok, 2004.
30. International Cooperation on Cosmetics Regulation. About ICCR. Available online: https://www.iccr-cosmetics.org/about-us [Last accessed on 14-Oct-2024].
31. Organization for Economic Co-Operation and Development (OECD). Recommendation of the Council Concerning Safety Controls over Cosmetics and Household Products. Available online: https://legalinstruments.oecd.org/en/instruments/OECD-LEGAL-0146 [Last accessed on 09-Oct-2024].
32. International Organization for Standardization (ISO). Technical Commitees, ISO/TC 217, Cosmetics. Available online: https://www.iso.org/committee/54974.html [Last accessed on 09- Oct-2024].
33. Ferreira, M., Matos, A., Couras, A., Marto, J., Ribeiro, H. Overview of Cosmetic Regulatory Frameworks around the World. *Cosmetics*, 2022, 9, 72.
34. EMA - Medicines - Search - European Union herbal monograph. Available from: https://www.ema.europa.eu/en/human-regulatory-overview/herbal-medicinal-products/european-union-monographs-list-entries. [Last accessed on 09-Oct-2024].
35. European Parliament and Council of the European Union. Regulation (EC) No 1223/2009 of the European Parliament and of the Council of 30 November 2009 on Cosmetic Products; 2009. Available from: https://eur-lex.europa.eu/legal-content/EN/ALL/?uri=celex%3A32009R1223 [Last accessed on 26-Sept-2024].
36. European Commission. Regulation 1223/2009 on Cosmetic Products, as Amended, *OJEU*, L342/59-209; 22.12.2009. https://health.ec.europa.eu/system/files/2016-11/cosmetic_1223_2009_regulation_en_0.pdf [Last accessed on 09-Oct-2024].
37. European Parliament and Council of the European Union. Regulation (EC) No. 1223/2009 of the European Parliament and of the Council of 30 November 2009 on Cosmetic Products; 2009. Available from: https://eur-lex.europa.eu/eli/reg/2009/1223/2024-04-24/eng [Last accessed on 27-Sept-2024].
38. European Union publications. Impact Assessment Study on Fragrance Labelling on Cosmetic Products. https://op.europa.eu/en/publication-detail/-/publication/71005243-433b-11eb-b27b-01aa75ed71a1/language-en [Last accessed on 27-Sept-2024].
39. https://ansm.sante.fr/uploads/2021/03/11/657257784ff10b16654e1ac94b60e3fb.pdf [Last accessed on 09-Oct-2024].
40. https://ansm.sante.fr/uploads/2021/03/11/599485607ae049abfac313b71943d612.pdf [Last accessed on 09-Oct-2024].
41. Bundesinstitut für die Risikobewertung (BfR). FAQs, 'Frequently Asked Questions about the Use of Essential Oils', 28.02.2008. https://www.bfr.bund.de/cm/349/frequently_asked_questions_about_the_use_of_essential_oils.pdf.
42. European Directorate for the Quality of Medicines & HealthCare of the Council of Europe (EDQM). *Guidance on essential oils in cosmetic products*, 2016. https://www.edqm.eu/en/guidance-on-essential-oils-in-cosmetic-products [Last accessed on 14-Oct-2024].
43. https://www.bfr.bund.de/cm/349/use_of_undiluted_tea_tree_oil_as_a_cosmetic.pdf [Last accessed on 09-Oct-2024].
44. Council of Europe. *Plants in Cosmetics: Plants and Herbal Preparations Used as Ingredients in Cosmetics*. Volume I. Patri, F., and Silano, V. 2001.
45. Council of Europe. *Plants in Cosmetics: Plants and Herbal Preparations Used as Ingredients in Cosmetics*. Volume II. Anton, R., Patri, F. and Silano, V. 2001.

46. Council of Europe. *Plants in Cosmetics: Potentially Harmful Components.* Volume III. Committee of Experts on Cosmetic Products. 2006.
47. Council of Europe. *Active Ingredients Used in Cosmetics: Safety Survey.* 2008.
48. Council of Europe. *Safe Cosmetics for Young Children. A Guide for Manufacturers and Safety Assessors.* ISBN 978-92-871-7337-9. 2023.
49. European Commission. Nanomaterials. Available online: https://single-market-economy.ec.europa.eu/sectors/cosmetics/cosmetic-products-specific-topics/nanomaterials_en [Last accessed on 28-Aug-2025].
50. Official Journal of the European Union. L 2024/858, 15 March 2024. http://data.europa.eu/eli/reg/2024/858/corrigendum/2024-07-31/oj [Last accessed on 14-Oct-2024].
51. https://single-market-economy.ec.europa.eu/sectors/cosmetics/ban-animal-testing_en [Last accessed on 09-Oct-2024].
52. US Government Publishing Office. Electronic Code of Federal Regulations. Title 21: Food and Drugs. Parts 700 to 799; 2025. https://www.ecfr.gov/current/title-21/chapter-I/subchapter-G [Last accessed on 28-Aug-2025].
53. Available from: https://www.fda.gov/cosmetics/cosmetics-laws-regulations/fda-authority-over-cosmetics-how-cosmetics-are-not-fda-approved-are-fda-regulated [Last accessed on 01-Oct-2024].
54. https://www.fda.gov/cosmetics/cosmetic-ingredients/fragrances-cosmetics#labelling [Last accessed on 30-Sept-2024].
55. https://www.ecfr.gov/current/title-21/chapter-I/subchapter-G/part-701 [Last accessed on 28-Aug-2025].
56. https://www.ecfr.gov/current/title-21/chapter-I/subchapter-G/part-701/subpart-A [Last accessed on 11-Oct-2024].
57. https://www.ecfr.gov/current/title-21/chapter-I/subchapter-G/part-701/subpart-B/section-701.13 [Last accessed on 11-Oct-2024].
58. https://www.ecfr.gov/current/title-21/chapter-I/subchapter-G/part-740 [Last accessed on 11-Oct-2024].
59. https://www.fda.gov/cosmetics/cosmetic-ingredient-names/color-additives-permitted-use-cosmetics [Last accessed on 28-Aug-2025].
60. https://www.fda.gov/media/88234/download [Last accessed on 01-Oct-2024].
61. https://www.fda.gov/media/86366/download [Last accessed on 28-Aug-2025].
62. https://www.fda.gov/media/86366/download [Last accessed on 01-Oct-2024].
63. https://www.fda.gov/food/cfsan-constituent-updates/fda-begins-posting-adverse-event-report-data-foods-and-cosmetics [Last accessed on 01-Oct-2024].
64. https://www.fda.gov/cosmetics/cosmetics-labeling/cosmetics-labeling-claims [Last accessed on 28-Aug-2025].
65. ISO 22716:2007 https://www.iso.org/obp/ui/en/#iso:std:iso:22716:ed-1:v2:en [Last accessed on 01-Oct-2024].
66. https://www.iso.org/committee/54974.html [Last accessed on 01-Oct-2024].
67. https://www.iso.org/standard/65197.html [Last accessed on 01-Oct-2024].
68. Es, I., Khaneghah, A. M., Akbariirad, H. Global Regulation of Essential Oils. In Essential Oils in Food Processing: Chemistry, Safety and Applications, Hashemi, S. M. B., Khaneghah, A. M., de Souza, A., Eds., Wiley-Blackwell: Hoboken, NJ, 2017, pp. 327–338.
69. https://ec.europa.eu/health/scientific_committees/opinions_layman/perfume-allergies/en/l-3/1-introduction.htm [Last accessed on 04-Oct-2024].
70. https://ifrafragrance.org/ [Last accessed on 01-Oct-2024].
71. https://ifrafragrance.org/about-ifra/ifra-code-of-practice [Last accessed on 01-Oct-2024].
72. SCCNFP. The Scientific Committee on Cosmetic Products and Non-Food Products Intended for Consumers, Fragrance Allergyin Consumers. A Review of the Problem. Analysis of the Need for Appropriate Consumer Information and

Identification of Consumer Allergens. 1999. Available online: https://ec.europa.eu/health/ph_risk/committees/sccp/documents/out98_en.pdf [Last accessed on 02-Sept-2024].
73. ASEAN Secretariat. ASEAN Cosmetic Directive, 2008. Available from: https://aseancosmetics.org/asean-cosmetics-directive/articles-of-acd/ [Last accessed on 04-Oct-2024].
74. https://aseancosmetics.org/uploads/UserFiles/File/PUBLICATIONS/Botanical_Safety_Assessment_Guidelines.pdf [Last accessed on 11-Oct-2024].
75. ASEAN Secretariat. ASEAN Cosmetic Directive, 2008.Available from: https://aseancosmetics.org/asean-cosmetics-directive/articles-of-acd/ [Last accessed on 11-Oct-2024].
76. The Drugs and Cosmetics Rules. Schedule M-II, 1945. Available from: https://cdsco.gov.in/opencms/opencms/en/Cosmetics/cosmetics/ [Last accessed on 11-Oct-2024].
77. The Drugs and Cosmetics Act, 1940. Available from: https://pib.gov.in/newsite/PrintRelease.aspx?relid=101617. [Last accessed on 11-Oct-2024].
78. https://ayush.gov.in/#!/quality_standard [Last accessed on 11-Oct-2024].
79. The Drugs and Cosmetics Rules. Schedule, 1945. Available from: https://ayush.gov.in/resources/pdf/quality_standards/Drugs-and-Cosmetics-Act-Rules.pdf [Last accessed on 11-Oct-2024].
80. National Medical Products Administration (NMPA). n.d. Available from: https://english.nmpa.gov.cn/cosmetics.html [Last accessed on 14-Oct-2024].
81. NMPA. n.d. Administrative Measures for the Registration and Filing of Cosmetics. Available from: https://english.nmpa.gov.cn/2021-03/05/c_661070.htm [Last accessed on 14-Oct-2024].
82. Provisions for Supervision and Administration of Manufacturing and Marketing of Cosmetics, 2022. https://english.nmpa.gov.cn/2022-10/25/c_824600.htm [Last accessed on 14-Oct-2024].
83. NMPA. Cosmetic Supervision and Administration Regulation (CSAR), 2020. Available from: https://english.nmpa.gov.cn/2020-07/03/c_506171.htm [Last accessed on 14-Oct-2024].
84. Ministry of Health, Labour and Welfare (MHLW). n.d. Available from: https://www.mhlw.go.jp/english/org/profile/index.html [Last accessed on 14-Oct-2024].
85. Ministry of Food and Drug Safety (MFDS). n.d. Available https://www.mfds.go.kr/eng/index.do [Last accessed on 07-Oct-2024].
86. National Industrial Chemicals Notification and Assessment Scheme (NICNAS). n.d. Available from: https://www.nicnas.gov.au/about-nicnas [Last accessed on 14-Oct-2024].
87. Australian Industrial Chemicals Introduction Scheme. Labelling, SDS and Packaging, 2022. https://www.industrialchemicals.gov.au/help-and-guides/labelling-sds-and-packaging [Last accessed on 14-Oct-2024].
88. https://www.industrialchemicals.gov.au/business/getting-started-registration-importing-and-manufacturing/organic-and-natural-ingredients [Last accessed on 14-Oct-2024].
89. Health Canada. Cosmetics and Personal Care. n.d. Available from: https://www.canada.ca/en/health-canada/services/consumer-product-safety/cosmetics/regulatory-information.html [Last accessed on 14-Oct-2024].
90. Brazilian Health Regulatory Agency (ANVISA). n.d. Available from: https://www.gov.br/anvisa/pt-br/assuntos/cosmeticos [Last accessed on 28-Aug-2025].
91. https://www.adipec.com.br/assets/pdf/ANVISA-DIRECTIVE-RDC-No7-10-FEBRUARY-2015.pdf. [Last accessed on 15-Oct-2024].
92. Federal Service for Surveillance in Healthcare (Roszdravnadzor). n.d. Available from: https://roszdravnadzor.gov.ru/en [Last accessed on 15-Oct-2024].

93. https://www.rustandard.com/images/CU_TR/TR_TS_009-2011.pdf [Last accessed on 15-Oct-2024].
94. International Cooperation on Cosmetics Regulation (ICCR). https://www.iccr-cosmetics.org/about-us. [Last accessed on 15-Oct-2024].
95. https://www.iccr-cosmetics.org/component/attachments/?task=download&id=171. [Last accessed on 15-Oct-2024].
96. https://www.cosmos-standard.org/en/cosmos-standard/ [Last accessed on 14-Oct-2024].
97. Personal Care Products Council (PCPC). n.d. About PCPC. Available from: https://www.personalcarecouncil.org/about-us/ [Last accessed on 14-Oct-2024].
98. European Cosmetic Industry Association (Cosmetics Europe). n.d. About Cosmetics Europe. Available from: https://cosmeticseurope.eu/about-us/ [Last accessed on 14-Oct-2024].
99. International Federation of Societies of Cosmetic Chemists (IFSCC). n.d. Available from: https://ifscc.org/ [Last accessed on 14-Oct-2024].
100. Society of Cosmetic Scientists (SCS). n.d. About SCS. https://scs.org.uk/ [Last accessed on 15-Oct-2024].
101. Personal Care Products Council (PCPC). n.d. International Nomenclature of Cosmetic Ingredients (INCI). Available from: https://www.personalcarecouncil.org/resources/inci/ [Last accessed on 14-Oct-2024].
102. The European Federation for Cosmetic Ingredients (EFfCI). n.d. About EFfCI. Available from: https://effci.com/index.php?p=sec-a#who-we-are [Last accessed on 14-Oct-2024].
103. The Natural and Organic Cosmetics Association (NATRUE). n.d. About NATRUE. https://natrue.org/who-we-are/ [Last accessed on 15-Oct-2024].
104. The Vegan Society. n.d. About The Vegan Society. Available from: https://www.vegansociety.com/about-us [Last accessed on 15-Oct-2024].
105. The Coalition for Consumer Information on Cosmetics (CCIC). n.d. About CCIC. Available from: https://www.leapingbunny.org/about/about-leaping-bunny [Last accessed on 14-Oct-2024].
106. Ecocert. n.d. About Ecocert. Available from: https://www.ecocert.com/en/about-ecocert [Last accessed on 14-Oct-2024].
107. Soil Association. n.d. About Soil Association Certification. Available from: https://www.soilassociation.org/certification [Last accessed on 14-Oct-2024].
108. Cosmetic Valley. n.d. About Cosmetic Valley. Available from: https://www.cosmetic-valley.com/en [Last accessed on 14-Oct-2024].
109. Cosmetics Alliance Canada. n.d. About Cosmetics, Alliance Canada. Available from: https://www.cosmeticsalliance.ca [Last accessed on 14-Oct-2024].
110. https://independentbeauty.org/ [Last accessed on 14-Oct-2024].
111. The Indian Beauty and Hygiene Association (IBHA). n.d. About IBHA. Available from: https://ibhaindia.com/ [Last accessed on 14-Oct-2024].
112. BDIH-Bundesverband der Industrie-und Handelsunternehmen. n.d. About BDIH. Available from: https://www-kontrollierte–naturkosmetik-de.translate.goog/bdih.htm?_x_tr_sl=de&_x_tr_tl=en&_x_tr_hl=en&_x_tr_pto=sc/ [Last accessed on 14-Oct-2024].
113. The Organic Materials Review Institute (OMRI). n.d. About OMRI. Available from: https://www.omri.org/about [Last accessed on 14-Oct-2024].
114. Cosmetic Ingredient Review (CIR). n.d. About CIR. Available from: https://www.cir-safety.org/about [Last accessed on 14-Oct-2024].
115. Green Chemistry and Commerce Council (GC3). n.d. About GC3. Available from: https://greenchemistryandcommerce.org/about-gc3/introduction [Last accessed on 14-Oct-2024].
116. Global Cosmetics Cluster (GCC). n.d. About GCC. Available from: https://www.cosmetics-clusters.com/ [Last accessed on 14-Oct-2024].

117. Cosmetic, Toiletry and Perfumery Association (CTPA). n.d. About CTPA. Available from: https://www.ctpa.org.uk/about-us [Last accessed on 14-Oct-2024].
118. The Australian Society of Cosmetic Chemists (ASCC). n.d. About ASCC. Available from: https://ascc.com.au/about-us/ [Last accessed on 28-Aug-2025].
119. Society of Cosmetic Chemists (SCC). n.d. About SCC. Available from: https://www.scconline.org/about [Last accessed on 14-Oct-2024].
120. Personal Care Products Council (PCPC). n.d. Science Symposium. Available from: https://www.personalcarecouncil.org/events/2025-summit/ [Last accessed on 14-Oct-2024].
121. Professional Beauty Association (PBA). n.d. About PBA. Available from: https://www.probeauty.org/ [Last accessed on 14-Oct-2024].
122. International Association for Soaps, Detergents and Maintenance Products (AISE). n.d. About AISE. Available from: https://aise.eu/ [Last accessed on 14-Oct-2024].

Chapter 7

Navigating regulatory differences and strategies for global compliance

Case study of multinational companies

Samreen Abbas, Saniya Shakir, Rajat Awasthi, Neha Nerma, and Bhartendu Nath Mishra
Department of Biotechnology, Institute of Engineering and Technology, Lucknow, India

Shiv Poojan
bioTOX Solution, NNF, IET, Lucknow, India

7.1 INTRODUCTION

Essential oils are likely useful as fragrances and cosmetics due to their volatile nature. Their intended usage has primarily been industrial for a long time, as standardizing fragrances by combining essential oils from several plants to get a certain fragrance has been a common process. Out of over 3000 plant species, only 300 essential oils are commercially available in the fragrance market. The essential oils from orange, corn mint, citronella, eucalyptus, peppermint, and lemon are the most commonly manufactured commercially. Because of their strong fragrance, natural marketing image, and complex active ingredients, essential oils are frequently used in today's skincare products (Coppen, 1994). They have also demonstrated several qualities applicable to cosmetic use that have been scientifically verified over time. For example, their value as a natural preservative agent in cosmetic preparations is proven by literature. Due to their antibacterial qualities, which essential oils can possess independently or in conjunction with other preservatives, they ensure protection against fungi and bacteria. Therefore, it is convenient that additional essential oils might improve the dermato-cosmetic qualities of the finished product by preventing microbial infections and aiding in the preservation of the cosmetic formulation (Bakkali et al., 2008). It's interesting to note that they are also utilized to promote other skin benefits like sun protection, skin whitening, anti-ageing, and anti-acne effects. The widespread use of essential oils is due to their fragrant and medicinal properties. Stringent regulatory procedures govern their usage in cosmetic products globally to ensure safety and compliance. Manufacturers must comprehend these regulations to navigate the complex regulatory procedure with effectiveness (Burger et al., 2019).

DOI: 10.1201/9781003558477-7

Fragrances play a particularly important role in increasing the attractiveness of cosmetics. Pleasant smells influence the comfort and effect of the products and also impact significantly the overall evaluation of cosmetics. Therefore, along with the shape and design of the container, the smell is also one of the characteristics of cosmetics that the consumers experience and look forward to in the selection of those items (Winkelman, 2018). Among natural fragrances, essential oils, which are complex mixtures of terpenes and other aromatic or aliphatic compounds, produced as secondary metabolites in specialized secretory tissues of aromatic plants, are the most popular. In nature, essential oils play very important roles in plant defence and signalling processes. For instance, they are involved in plant defence against insects, herbivores, and microorganisms, including attraction of pollinating insects and fruit-dispersing animals, water regulation, and allelopathic interactions (Wongsukkasem et al., 2018).

Exploration focuses on the sustainable business practices of firms that deal in medicinal aromatic herbs and essential oils, which are subject to resource depletion as a result of overharvesting and environmental issues. It draws attention to threatened species such as rosewood, frankincense, sandalwood, and myrrh, emphasizing the need for organic farming methods and indebted production. The data show that the majority of businesses, especially in Turkey, solely concentrate on packaging and resource usage, with very few addressing sustainability. The investigation emphasizes the necessity of ongoing, inclusive efforts to guarantee the long-term survival of these priceless resources and advocates for more emphasis on sustainable harvesting and industry awareness (Babita et al., 2014).

Only 9 out of 25 national companies in the Turkish essential oil market have a dedicated sustainability tab on their websites, based on an analysis. The following companies strongly emphasize natural and organic products: Bade Natural, Bodivi, Edy's Organics, Goya Botanicals, Misbahçe, Raqun, Rosece, Simya Evi, and the Elite Home. Though natural components are the main focus, only a small number of companies—namely, Goya Botanicals, Misbahçe, and Alfheim Aromatherapy—address sustainable methods including using local resources and conducting ethical harvesting. Meditolia and Rosemary are two businesses that don't have an internet presence and rely solely on e-commerce sites to make sales. Most businesses emphasize how their products are natural and additive-free and don't use chemicals or animal testing. Furthermore, not many discuss the crucial part of sustainable harvesting practices, essential for the long-term viability of medicinal plants (Essential Oils Market Size, Share & Growth Report, 2030, n.d.).

Other companies that highlight sustainability include Aethos, Aura Cacia, Aveda, doTERRA, Eden's Garden, Helias, and Jurlique. They focus on ethical sourcing, sustainable farming, and harvesting methods. Aveda has a strong emphasis on minimizing packaging waste, renewable energy, and responsible packaging. These companies also emphasize that they don't use chemicals or do animal testing, which is in line with the growing consumer desire for ecological and ethical products. Sustainable harvesting and production methods are crucial for the long-term viability of medicinal plants, even though packaging and renewable resources are also significant—a factor that not all businesses fully address. In cosmetics, fragrances are vital because they cover up smells that are unpleasant and make products look more appealing (Fonseca-Santos et al., 2015).

The increasing desire for natural components in cosmetics and health products has led to a surge in the popularity of essential oils, which are prized for their inherent fragrance, ability to preserve, and benefits for the skin. Essential oils such as citrus, lavender, and eucalyptus are frequently utilized, in addition to constituents like citral and linalool (Ben Hsouna et al., 2017). The demand for essential oils is driven by the food & beverage, cosmetics, and aromatherapy sectors worldwide, with Europe holding the largest market share. Large-scale production of common oils such as orange, lemon, and maize mint is done. Rare oils, such as agarwood and rose, are very expensive. The market for essential oils in the United States has grown dramatically, and by 2024, it was predicted to reach $7.3 billion. France continues to be the world's largest exporter, but important importers, particularly of citrus oils, include Northern Europe and Lebanon (Ibrahim et al., 2024).

Europe holds the largest market share for essential oils worldwide, with North America and Asia-Pacific region sharing second position. Scent-infused candles, oils, and other items are largely made with essential oils. The market value of the essential oils sector in the United States increased by about 10% in 2017 and was projected to reach $7.3 billion by 2024. The essential oils that account for a major portion of the US market are eucalyptus, maize mint, and orange oils. Moreover, France is the world's largest exporter of essential oils and has a tradition of producing perfumes. In recent years, the United States has exported approximately $2 billion dollars worth of essential oils and resinoids (Tripathi et al., 2016). Furthermore, some nations are forced to import specific essential oils from other nations because they are unable to produce them in enough quantities to meet domestic demand. For example, in 2017, Lebanon bought the greatest amount of essential oils of any nation in the world (around $13.75 billion), with nearly 80% of the orange oil imported into Europe coming from Northern European nations including Germany, the Netherlands, and the United Kingdom (Sharmeen et al., 2021).

Any essential oil or product containing essential oils meant for cosmetic use has to abide by certain regulations under Regulation 1223/09 on cosmetic products. This regulation defines a cosmetic product as "any substance or mixture intended to be applied to the external surface of the human body like hair, lips, nails, epidermis, hair system or mucous membrane of the mouth, teeth with the main purpose of cleaning, perfuming, changing their appearance, protecting them, keeping them in good condition, or correcting body odours" and protecting the epidermis, hair system, hair, nails, lips, and external genital organs (Essential Oil Regulations in the EU: A Guide, n.d.). An essential oil meant for cosmetic application needs to come from plants that are permitted to be used in these types of products. Ninety-two plant species and their derivatives are specifically listed as being prohibited for use in cosmetics in Annex II of the Cosmetics Regulation. Furthermore, there are strict regulations regarding the amount of some naturally occurring compounds found in essential oils, such as methyl eugenol, furocoumarins, and safrole, in final products (Sharifi-Rad et al., 2017). Cosmetics have to be safe to apply on the skin or skin appendages, meant only for external use, and not meant for internal consumption. They can be used for cleaning, scenting, altering appearance, safeguarding, preserving condition, or removing body odour. These roles may be primary or secondary as long as they adhere to Regulation (EC) 655/2013, which creates uniform standards for the validation of claims made in cosmetic goods. Table 7.1 shows a list of plant species from which essential oils are obtained as well as the aroma description used in different applications (Sharmeen et al., 2021).

Table 7.1 Aroma profiles of essential oils used as fragrances in cosmetics and perfumes

Sr no.	*Plant species from which essential oils are derived*	*Common name*	*Part(s) used*	*Color range of oils*	*Aroma description*
1	*Amyris baisamifera* L	Balsam torchwood	Wood	Pale yellow to amber	Characteristic, woody
2	*Angelica archangelica*	Angelica	Roots and rhizomes	Colourless or pale yellow	Rich herbaceous-earthy body note
3	*Aniba rosaeodora* Ducke	Rosewood	Wood/leaves	Pale yellow	Floral, sweet, woody, and citric
4	*Boswellia sacra* Flueck	Frankincense	Gum	Clear or yellow	Balsamic, camphor-like, spicy, woody, slightly lemony
5	*Cananga odorata* Hook. F. and Thoms	Ylang-ylang	Flowers	Pale yellow to dark yellow	Characteristic, floral, recalling jasmine
6	*Carum carvi* L.	Caraway	Ripe fruit	Colourless to pale yellow	Fresh, herbaceous, spicy
7	*Cedrus atlantica* (Endl.) Manetti ex Carrière	Cedarwood	Wood/ sawdust	Yellowish to orange yellow or deep amber-coloured	Camphoraceous—cresylic top note with a sweet, tenacious woody undertone, reminiscent of cassie and mimosa
8	*Cinnamomum verum* J. Presl	Cinnamon	Bark	Reddish-brown	Characteristically spicy burning
9	*Citrus aurantiifolia* (Christm.) Swingle	Neroli/Bitter orange	Flowers	Pale yellow to coffee brown	Mild citrus, floral
10	*Citrus aurantium var. amara* L.	Neroli/Bitter orange	Flowers	Pale yellow to coffee brown	Sweet, fresh, and floral odour

(*Continued*)

Table 7.1 (Continued) Aroma profiles of essential oils used as fragrances in cosmetics and perfumes

Sr no.	*Plant species from which essential oils are derived*	*Common name*	*Part(s) used*	*Color range of oils*	*Aroma description*
11	*Citrus limon* (L.) Osbeck	Lemon	Peel	Colourless to pale yellow	Fresh lemon peel
12	*Citrus bergamia* Risso	Bergamot	Rind of fruit	Light green yellow	Light, delicate citrus-like scent with slightly floral/ sweet/balsamic character
13	*Citrus reticulata* Blanco	Tangerine	Peel	Dark orange to reddish-orange or brownish-orange	Orange-like
14	*Citrus sinensis* (L.) Osbeck	Sweet orange	Peel	Yellow to reddish-yellow	Characteristic, orange peel odour
15	*Citrus paradisi* Macfad	Grapefruit	Peel of fruits	Yellow or greenish	Fresh, sharp, citrus aroma
16	*Commiphora myrrha* (Nees) Engl.	Myrrh	Gum	Yellow to reddish-brown	Pleasant balsamic, camphor-like, musty, incense-like
17	*Coriandrum sativum* L.	Coriander	Fruits	Colourless to pale yellow	Floral, balsamic undertone, and peppery-woody suave
18	*Cupressus sempervivens* L.	Cypress	Leaves	Yellowish	Pleasingly smoky, woody smell, and amber-like
19	*Cymbopogon citratus* (DC.) Stapf	Lemongrass	Grass leafs	Yellow or pale sherry-coloured	Intense sweet lemony, reminiscent of lemon drops
20	*Cymbopogon martini* (Roxb.) W.Watson	Palmarosa	Leaves	Yellow	Fresh rose-like
21	*Daucus carota* L.	Carrot seed	Seeds	Yellowish-brown	Soft, sweet earthy grounding aroma
22	*Elettaria cardamomum* (L.) Maton	Cardamom	Fruits	Almost colourless to pale yellow	Characteristic, spicy, cineolic
23	*Eucalyptus globulus* Labill.	Eucalyptus	Leaves	Yellow to red	Strong, anisey, camphoric
24	*Feniculum vulgare* Mill.	Fennel	Seeds	Yellowish	Strong, anisey, camphoric
25	*Helichrysum italicum* (Roth) G. Don	Immortelle	Aerial parts	Pale yellow to red	Strong, honey-like aroma

7.2 GLOBAL REGULATORY PRACTICES

The presence of organizations such as the World Health Organization (WHO), Food and Agriculture Organization (FAO) of the United Nations, International Organization for Standardization (ISO), and Food and Drug Administration (FDA) has already helped the essential oil market to be well controlled at the global level. EU Commission and regulatory authorities in the EU consider essential oils' biologically active substances. The Scientific Committee of the Essential Oil Association (EOA) of the United States, between 1940 and 1975, showed intense activities in the United States to develop specifications and test methods of essential oils. Like other organizations, Australian food law agencies established some legislation about essential oils. Figure 7.1 depicts the import values of leading countries in the world for essential oils and retinoids.

7.2.1 United States

The FDA considers essential oils as plant ingredients that might contain allergens or other harmful substances when used in applications such as cosmetics. As such, essential oils are regulated under FDA 21 CFR, California Proposition 65, and other regulations (Essential Oil Regulations in the United States: An overview, n.d.).

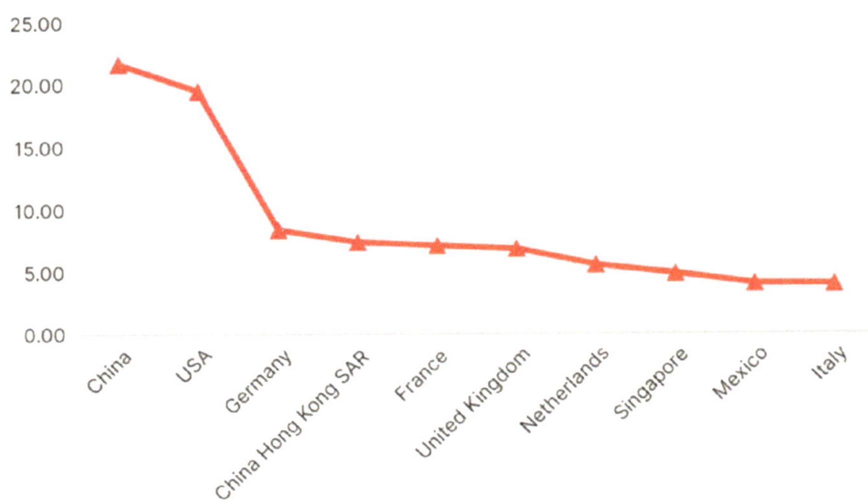

Figure 7.1 Import value of essential oils and resinoids worldwide in 2021, by leading country. (Adapted and modified from Essential Oils and Resinoids: Top Importers Worldwide 2021.)

7.2.1.1 FDA 21 CFR

Among various other products, the FDA 21 CFR governs food, medicines, and cosmetics. Certain restrictions apply to essential oils used in food and cosmetics, among other product categories.

7.2.1.1.1 Requirements

The FDA requires importers and manufacturers to:

a. Properly label their cosmetics, and
b. Avoid misbranding, and providing false information with, their cosmetics.

It also recommends having products tested to ensure product safety. However, the FDA does not:

a. Require cosmetics to have pre-market approval, except when they contain colour additives.
b. Require manufacturers to register cosmetic establishments or product formulations, although it encourages participation in the FDA's Voluntary Cosmetic Registration Program (VCRP).

7.2.1.1.2 Drugs

The FDA classifies essential oils and other goods intended for therapeutic purposes or to alter physiological states as medicines. For example, the FDA considers drugs as any product made with essential oils and marketed as having the ability to cure pain or anxiety. Furthermore, essential oils are frequently included in products made for aromatherapy. If these items were intended for the treatment or prevention of illness, the FDA would classify them as drugs. If massage oils and lotions are such that they can relieve pain, then these goods may also fall into this category (ECFR: 21 CFR Part 700 Subpart A – General Provisions, n.d.).

7.2.1.2 The Toxic Substances Control Act (TSCA)

The Toxic Substances Control Act (TSCA) sets requirements for chemical substances, including substances that might be found in essential oils.

7.2.1.2.1 TSCA inventory

The TSCA Inventory contains over 83,000 chemicals and lists substances that might be used in essential oils, by names and definitions (e.g., extractives and their physically modified derivatives) (Rayasam et al., 2022).

7.2.1.3 The Federal Hazardous Substances Act (FHSA)

The Federal Hazardous Substances Act (FHSA) provides labelling requirements regarding hazardous substances. The Act also covers strong sensitizers, which can cover some essential oils.

7.2.1.3.1 Sensitizers and strong sensitizers

Sensitizers are defined as compounds that, upon exposure, may cause the immune system to become hypersensitive to them (16 CFR Part 1500.3). An allergic reaction occurs as a result of the hypersensitivity after being exposed to the sensitizer repeatedly. Conversely, strong sensitizers are compounds that, when applied again, result in allergic hypersensitivity.

7.2.1.3.2 Requirements

As per the Federal Hazardous Substances Act (FHSA) Requirements | CPSC. Gov (n.d.), the FHSA requires packaging for hazardous substances to include the following information on the label:

- Importer's and manufacturer's name and business address
- The common or chemical name of each hazardous ingredient
- "Danger" is the signal word for products that are hazardous, combustible, or corrosive
- The warning symbol "Caution" or "Warning" for further dangerous goods
- A statement of the product's main hazard, e.g., "Flammable"
- Precautionary statements
- First-aid instructions when applicable
- The warning labels "Danger" and "Poison" for extremely hazardous goods
- Protection instructions for products needing special handling care
- The statement "Keep out of the reach of children"

7.2.1.4 The Fair Packaging and Labelling Act (FLPA)

The Fair Packaging and Labeling Act (FPLA) mandates labelling requirements for "consumer commodities".

7.2.1.4.1 Requirements

Products, which may include essential oils, are required by the FPLA to have labels with the following information on them:

a. The product's identity
b. The importer or manufacturer's name and business address

c. The net quantity of contents, stated in both inch/pound and metric units (in terms of mass or weight, measure, or numerical count) (Fair Packaging and Labeling Act: Regulations Under Section 4 of the Fair Packaging and Labeling Act | (Federal Trade Commission, n.d.)

7.2.1.5 California Proposition 65

Certain compounds, listed on the Proposition 65 list and their usage in consumer products are restricted by California Proposition 65. Some of those prohibited substances might be present in essential oils. Products that contain restricted chemicals over permitted levels must have a warning label. Certain naturally occurring compounds that could be used as constituents in essential oils are included in the Proposition 65 list (Guirguis et al., 2014).

The two substances listed below are included because they have the potential to cause cancer:

- Beta-myrcene (used for aroma, flavouring, and fragrance purposes)
- Pulegone (used for flavouring, fragrance, and herbal purposes)

7.2.1.6 ISO Standards

Essential oil standards have been released by the International Organization for Standardization (ISO). While each of those standards has a distinct function, some of them help determine the qualities of the oil, which helps with quality assessments.

Here are a few examples of such standards:

a. ISO/TR 21092—Essential oils—Characterization
b. ISO 4715—Essential oils—Quantitative evaluation of residue on evaporation
c. ISO 11021—Essential oils—Determination of water content—Karl Fischer method

7.2.1.7 Lab testing

Manufacturers and importers are responsible for making sure that their essential oil products meet all applicable regulations (such as those of ingredients and packaging). Additionally, they have to make sure that the norms and laws that apply to their essential oil products are followed in terms of quality. To accomplish this, they can submit their products for laboratory testing to demonstrate compliance with relevant regulations, such as those described in the TSCA, FHSA, or PPPA. They get a test report proving compliance with appropriate regulations when their essential oil goods pass testing (ISO/TC 54 - Essential Oils, n.d.).

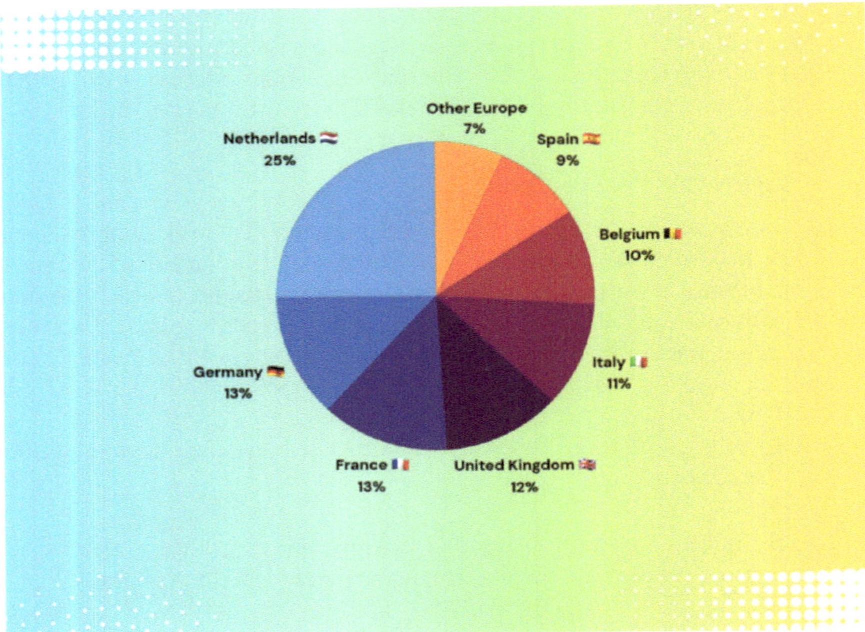

Figure 7.2 Leading European exporters of essential oils (2016, value).

7.2.2 European Union legislation for essential oils

The terms "general regulations" (also known as REACH and CLP) and "specific regulations" relate to the relevant legislative requirements regarding essential oils. General aspects of the main rules governing essential oils and the specific product categories (Figure 7.2).

7.2.2.1 REACH

REACH defines procedures for collecting and assessing information about the properties and hazards of chemical substances. Companies need to register their substances, and to do this, they need to work together with other companies that are registering the same substance. REACH only applies to European companies. Non-EU companies are not subject to its requirements, even if they export their products to the European Union (EU) (Barbieri & Borsotto, 2018). A company that wants to import more than 1 tonne of essential oil per year must either register its imports under REACH or buy from a company that does have REACH registration. That could be your company if you have set up an Only Representative. And if your registration covers the volumes imported each year by your customer.

Another important aspect of REACH is that it contains requirements for the Safety Data Sheet (SDS). All essential oils need an SDS when shipped to and within Europe; so you must provide one. Because essential oil is a complex mixture, its SDS must include hazardous information for its main components as well as for the product as a whole. The classification, labelling, and packaging (CLP) is based on the United Nations' Globally Harmonized System of Classification and Labelling of Chemicals (GHS). Many countries have made this part of their law. The CLP's requirements will therefore be familiar to non-EU companies that need to classify, label, package, and transport chemicals in their own countries. CLP data are included on the Safety Data Sheet (Cosmetics Regulation and REACH-Questions and Answers on Animal Testing What Does REACH Say about Animal Testing? n.d.).

7.2.2.2 Technical documentation

To comply with the EU's legal requirements, European buyers of essential oils for cosmetics need you to provide a well-prepared technical dossier. This not only demonstrates your commitment to quality and reliability but also your concern for sustainable and ethical sourcing

The technical dossier should include the following items:

- A technical data sheet (TDS)
- An SDS
- A certificate of analysis (CoA)
- An allergen declaration
- Information on traceability back to origin
- Responsible sourcing policies and practices
- Records of certifications awarded, and standards applied

7.2.2.3 Convention on International Trade in Endangered Species (CITES)

To export essential oils to Europe, you must comply with the rules on trading plant resources agreed internationally under CITES, the Convention on International Trade in Endangered Species of Wild Fauna and Flora. This is because the EU has translated CITES into EU law under Regulation 338/97. It is also likely your own country is a signatory of CITES, meaning that your national laws require you to comply with the convention as well (Raymakers, 2006).

CITES aims to ensure that international trade in wild animals and plants does not threaten the survival of rare species. The species covered are listed in three appendices to the convention.

- Appendix I includes species threatened with extinction. Trade in these species is permitted only in exceptional circumstances.

- Appendix II includes species not necessarily threatened with extinction, but in which trade must be controlled to avoid utilization incompatible with their survival.
- Appendix III contains species that are protected in at least one country, which has asked other CITES parties for assistance in controlling their trade.

7.2.2.4 Labelling and packaging

Besides complying with the CLP rules set out in EU Regulation 1272/2008, consider meeting other common labelling and packaging requirements that European buyers have.

European buyers require good-quality essential oils; so consider the following packaging tips to preserve the quality of your products in transit.

- Use aluminium, lined, or lacquered steel containers that conform to UN standards, as they do not react with the components of essential oils.
- Make sure that containers such as drums are perfectly clean and dry before filling them with essential oils.
- Fill the headspace of containers with gases that do not react with components of essential oils—carbon dioxide or nitrogen, for instance.
- Make sure that container sizes match the order volume. For example, 25 L in a 25 L container.

Other ways to preserve quality include making sure that your essential oils are kept at the right temperature throughout the supply chain and always storing them in a dry place.

7.2.2.5 Perfumers and aroma therapists

The third channel into the EU market is small-scale perfumers and aroma therapists. While the quantities they buy are smaller than importer-distributors or manufacturers want, they can be interested in ordering directly. There are two reasons for this. Firstly, essential oils are used directly (without further processing) in aromatherapy and some can be used directly in perfumes. Secondly, the small quantities required mean that their importation is not covered by REACH. The main challenge with this direct trade is international shipping. Due to essential oils' status as hazardous goods, special precautions need to be taken with their packaging and transport. This adds to the cost of shipping (Entering the European Market for Essential Oils | CBI, n.d.).

7.2.3 China

7.2.3.1 CSAR

The National Medical Products Administration (NMPA) is stepping up its enforcement of the Cosmetics Supervision and Administration Regulations (CSAR) China cosmetics standards. As demonstrated by the company Jian Shan Lake's recent decision to halt manufacturing of lavender essential oil, there is a growing number of violations of the CSAR. The Anhui Medical Products Administration (MPA) was notified by China's NMPA on April 15, 2022, that Jian Shan Lake's manufacturing of lavender essential oil has been suspended (No.1-3-2022). The essential oil's 2.5% "vanilla butyl ether" content in the product recipe, which is more than the 1.5% permitted by the Inventory of Existing Cosmetic Ingredients in China (IECIC) 2021, is the cause of the suspension.

Above 1.5% is regarded as dangerous for human health and poses a safety risk. When production poses a risk to human health or causes injury to others, the CSAR may be used to enforce a halt in production. "All business operations and production related to the cosmetics product, either causing human injury or with evidence indicating potential health risks on human beings, can be suspended immediately by the department in charge under the NMPA," according to Article 54 of the CSAR (China's New Cosmetics Regulation (CSAR), n.d.).

7.2.3.2 CSAR: China cosmetics infringements avoidance

To prevent the manufacturing, marketing, or sale of cosmetics in China from being suspended, compliance with the CSAR regulations on product formulas is essential. Since the product formula is an essential part of filing and registering a cosmetic product in China, applicants should make sure the content of raw materials used in cosmetics complies with the Chinese legislation. If a manufacturer faces suspension, they have 60 days to request administrative reconsideration from the NMPA or the relevant MPA. An additional option is to bring an administrative case before the local court within six months. Cinema can help in making sure that the business complies with China's most recent regulations regarding product quality and safety. From registration to post-market compliance to in-China OEM and subsidiary localization, cinema takes care of all needs (CSAR: China's New Cosmetic Regulation and What It Means, n.d.).

7.2.4 Japan

7.2.4.1 Regulation and procedure

7.2.4.1.1 The Foreign Exchange and Foreign Trade Act

Certain endangered plant species cannot be imported, as per the provisions of the Washington Convention on Foreign Exchange and Trade. The CITES laws apply to an essential oil of musk that is extracted from the abdomen of musk deer. Importing and exporting some types of rosewood, such as Brazilian Rosewood, is forbidden under CITES regulations (Japan Cosmetic Regulation I ChemLinked, n.d.).

7.2.4.2 The Pharmaceutical Affairs Act

When importing essential oils as drugs, quasi-drugs, or raw materials, the importer is required by the revised Pharmaceutical Affairs Act, which took full effect on June 1, 2006, to obtain a primary distributor's license for drugs (or quasi-drugs) from the prefectural governor who manages the business office where the marketing supervisor-general is employed. Furthermore, for each product item imported, the importer is required to seek primary distribution approval from the Minister of Health, Labour and Welfare (MHLW) (hereinafter referred to as "the Minister").

In this context, however, approval is not needed for drugs that are used exclusively in manufacturing and listed in the Japan Pharmaceutical Codex or the Japan Crude Drug Codex, or those listed in the Japanese Pharmacopoeia, an official standard reference that lists the names of significant drugs and defines quality standards of these drugs. For every item, the primary distributor can import without authorization.

The classification of importer licenses was removed by the amended Act. To package, label, or store an imported product, a major distributor must independently get a manufacturer's license for pharmaceuticals (or quasi-drugs). For the duration of these licenses, those who earned an importer's license under the previous legislation will be considered holders of a primary distributor's license and a manufacturer's license as a transitional step to the updated Act's provision (Japan External Trade Organization (JETRO) Development Cooperation Division Trade and Economic Cooperation Department, 2011).

7.2.4.3 Labelling

The Pharmaceutical Affairs Act mandates that the direct container, packaging, or package inserts be labelled with the relevant things when selling essential oils as cosmetics or quasi-drugs to guarantee appropriate consumption and handling, ensure quality, and establish culpability. All requirements must be stated in Japanese and must be specifically and unambiguously listed. In theory, all ingredient names for cosmetics have to be stated on the label. Names must generally be listed in a discernible order based on quantity and must appear in easily understood Japanese.

To be used by the Pharmaceutical Affairs Act's need, all ingredient names be listed on labels, the Japan Cosmetic Industry Association has put out a Japanese version of the "List of Cosmetic Ingredient Label Names." It is permitted to refer to them as "fragrance" when utilizing them as a fragrance agent.

- Brand name; contents (weight, volume, or number of units) (for pharmaceuticals and quasi-drugs)
- Name and address of major distributor; code or manufacturing number contents
- Ingredients list (generally, all ingredient names have to be given for cosmetics)
- Other goods specified by MHLW Ministerial Ordinance
- Expiration date, for those designated by the Minister.

The Aroma Environment Association of Japan (AEAJ) has implemented a labelling qualification programme for essential oils, which sets basic information and precautions for use to be labelled and standards for labelling, to provide information that helps consumers choose essential oils suitable for use in aromatherapy. The efficacy, safety, and other aspects of product quality have been standardized by the Air Fresheners & Deodorizers Conference.

It requires that the product name, application, composition, and content be labelled on the box, along with details on the manufacturer number, how to use it, safety measures, manufacturer or importer's name, address etc. Products that satisfy the Conference's requirements may be marketed with the compliance mark (Inoue & Hayashi, 2023).

7.2.5 India

India has long been known for its exquisite fragrances; essential oil perfumes were formerly part of the ancient royal wisdom, and the country has long practised aromatherapy, incense, and altar. The first scents utilized were the therapeutic scents brought about by Ayurveda, which is still widely recognized today and advocated using aromatic herbs and plants for age control, mental health, beauty, and medicinal purposes. The Indian fragrance business is one of the largest in terms of production and consumption, and at present, the fragrance market is poised to expand and give endless prospects for new players to thrive in this market. With its 15 different climatic zones, India is the world's most diverse country in terms of both flora and fauna. India can provide fragrant organic raw ingredients that are in high demand worldwide. India provides about $500 million to the $24.10 billion global fragrance and flavour sector. Though it has grown at a rate of about 11% over the past few years, India is expected to grow exponentially in the years to come due to rising demand from the middle class, rising personal care, rising brand awareness, rising disposable income, and affordable fragrance in the form of mass perfumes and deodorants. India is a global leader with nearly 90% of the total global production with nearly 40,000 tonnes of oil. Uttar Pradesh itself gives 80% of the total produce of India while Bihar gives 13%

and Madhya Pradesh gives 7% (Do I Need a License to Sell Essential Oils in India - Oil Therapy Central, n.d.).

7.2.5.1 Central Drugs Standard Control Organization (CDSCO)

Several government organizations, notably the Central Drugs Standard Control Organization (CDSCO) and the Bureau of Indian Standards (BIS), are in charge of regulating essential oils and fragrances in India. Since essential oils are frequently used to prevent germs and fungi in food and drinks, manufacturing and selling them requires a license from the Food Safety and Standards Authority of India (FSSAI) (CDSCO, n.d.).

7.2.5.1.1 Fragrance & Flavour Development Centre

The Government of India established the Fragrance & Flavour Development Centre (FFDC) in 1991 with support from the Government of Uttar Pradesh and UNDP/UNIDO. Technology has been provided by UNDP/UNIDO, land, buildings, and infrastructure by the Government of Uttar Pradesh, and indigenous, recurring, and non-recurring apparatus and equipment by the Indian government. The centre was established with the idea of serving as a conduit or interface between the industries and research surrounding essential oils, fragrances, and flavours. Selling essential oils for personal use in India does not require a specific license, as long as the oils are not marketed for commercial purposes. Individuals can use essential oils in their personal care routines and for aromatherapy without the need for commercial licensing (Figure 7.3) (FFDC Kannauj | Kannauj Attar, n.d.).

7.2.5.2 Food Safety and Standards Authority of India (FSSAI) Registration

7.2.5.2.1 Registering

FSSAI registration confirms compliance with legal requirements and food safety standards. Application must be submitted along with all necessary supporting documentation, including evidence of identification, proof of address, and a list of the items you plan to offer, to apply for FSSAI registration as an essential oil vendor in India. After that, FSSAI assesses the application to make sure that the production, distribution, and storage of essential oils adhere to their strict guidelines. Upon approval, FSSAI will provide you with a license (Apply for FSSAI Food Licenses Online | FoodLicenceApply, n.d.).

7.2.5.2.2 Import and export license

For Indian companies that deal in essential oils and are involved in international trade, obtaining import and export permits is crucial to enabling

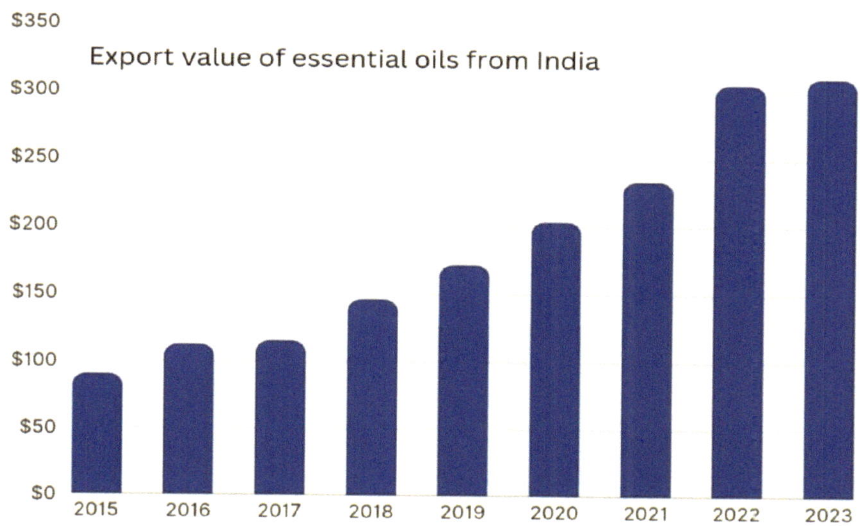

Figure 7.3 Export value of essential oils from India from the financial year 2015 to 2023. (Adapted and modified from Turek & Stintzing, 2013.)

cross-border operations by trade laws and customs rules. Businesses have to understand and adhere to the rules established by the Customs Department and the Directorate General of Foreign Trade (DGFT) before beginning to import or export essential oils into India.

7.2.5.2.3 Trade license

The acquisition of a commercial license is a key step in proving the validity of an essential oil enterprise in India. These permits are issued by municipal authorities following a careful investigation to confirm the company complies with local laws. Businesses can operate within a certain geographic area and comply with the law by obtaining a trade permit, which grants them the necessary legal recognition to carry out their operations.

7.2.5.2.4 Safety regulations

Strict safety rules must be followed when selling essential oils in India to protect consumers and maintain product quality. Adhering to international quality benchmarks, labelling criteria, and particular dilution protocols are all part of complying with safety regulations. Requirements for labelling and packaging are crucial for informing customers about critical information such as usage guidelines, possible allergies, and expiration dates. Products are guaranteed to meet quality and safety requirements when they comply with international safety standards, such as those established by the ISO.

7.2.5.2.5 Proper labelling and packaging

- In addition to assisting consumers in recognizing the product, good labelling conveys important information about components, usage guidelines, and storage suggestions.
- Important details like batch numbers, dates of expiration, and certifications ought to be visible on the labels.
- Essential oils require proper packaging to maintain their potency and integrity since it shields them from environmental elements like light and moisture.
- Adherence to safety regulations in labelling and packaging is essential to fulfil legal obligations and guarantee the welfare of consumers.

7.2.6 International Fragrance Association (IFRA)

Essential oils are regulated differently in different countries, and companies that export essential oils must adhere to international regulations. The International Fragrance Association (IFRA) and the EU have their regulations for essential oils. The IFRA provides guidelines on the safe use of fragrances in products, while the EU has regulations for labelling, packaging, and the safety of essential oils (Table 7.2) (Introduction, n.d.).

Table 7.2 The effect of essential oils on skin with their critical effects and recommendation

Sr no.	*Essential oils*	*Recommendation*	*Effect on skin*	*Critical effect*
1.	Acetylated vetiver oil	Restricted	Negligible skin contact	Dermal sensitization
2.	Alantroot oil	Prohibited	N/a	Sensitization
3.	Angelica root oil	Restricted	Leave-on products: 0.8% Non-skin contact products: no restriction	Phototoxicity
4.	Bergamot oil expressed	Restricted	Leave-on products: 0.4% Non-skin contact products: no restriction	Phototoxicity
5.	Bitter orange peel oil expressed	Restricted	N/a	Phototoxicity
6.	Boldo oil	Prohibited	N/a	Insufficient data
7.	Cade oil	Specification / prohibited	N/a	Carcinogenicity, genotoxicity*

(Continued)

Table 7.2 (Continued) The effect of essential oils on skin with their critical effects and recommendation

Sr no.	*Essential oils*	*Recommendation*	*Effect on skin*	*Critical effect*
8.	Chenopodium oil	Prohibited	N/a	The material had been in the list of "other materials" before which, for reasons of a unified standard format for prohibited materials, was disbanded.
9.	Citrus oils and other furocoumarins containing essential oils	For applications on areas of skin exposed to sunshine, excluding bath preparations, soaps, and other products which are washed off the skin	Leave-on products: 15 ppm 5-mop Non-skin contact products: no restriction	Phototoxicity
10.	Costus root oil, absolute and concrete	Prohibited	N/a	Sensitization
11.	Cumin oil	Restricted	N/a	Phototoxicity
12.	Fig leaf	prohibited		sensitization, phototoxicity
13.	Grapefruit oil expressed	Restricted	Leave on products: 4% Non-skin contact products: no restriction	Phototoxicity
14.	Jasmine absolute	There is no justification for a restriction of the concentration of this fragrance ingredient in the finished product	N/a	Sensitization
15.	Lemon oil cold pressed	Restricted	Leave on products: 2% Non skin contact products: no restriction	Phototoxicity
16.	Lime oil expressed	Restricted	Leave on products: 0.7% Non skin contact products: no restriction	Phototoxicity

(*Continued*)

Table 7.2 (Continued) The effect of essential oils on skin with their critical effects and recommendation

Sr no.	*Essential oils*	*Recommendation*	*Effect on skin*	*Critical effect*
17.	Massoia bark oil	Prohibited	N/a	The material had been on the list of "other materials" before which, for reasons of a unified standard format for prohibited materials, was disbanded.
18.	Melissa oil	Restricted	Due to the negligible skin contact from these types of products, there is no justification for a restriction of the concentration of this fragrance ingredient in the finished product	Sensitization
19.	Musk ambrette	Prohibited	N/a	Photosensitization, neurotoxicity
20.	Oakmoss extracts	Restricted	The level of oak moss in the respective category has to be reduced accordingly such that the total amount of both extracts does not exceed the maximum permitted level	Sensitization
21	Peru balsam crude	Prohibited	Restrictions on the use of peru balsam oil, absolute and anhydrol are contained in a separate standard (peru balsam extracts and distillates)	Sensitization
22.	Peru balsam extracts and distillates	Restricted	Due to the negligible skin contact from these types of products, there is no justification for a restriction of the concentration of this fragrance ingredient in the finished product	Sensitization
23.	Rue oil	Restricted	Leave on products: 0.15% Non skin contact products: no restriction	Phototoxicity

(*Continued*)

Table 7.2 (Continued) The effect of essential oils on skin with their critical effects and recommendation

Sr no.	Essential oils	Recommendation	Effect on skin	Critical effect
24.	Santolina oil	Prohibited	N/a	The material had been on the list of "other materials" before which, for reasons of a unified standard format for prohibited materials, was disbanded.
25.	Savin oil	Prohibited / specification	N/a	N/a
26.	Tagetes oil and absolute	Restricted	Leave-on products: 0.01% Non-skin contact products: no restriction	Phototoxicity
27.	Tea leaf absolute	Restricted	Due to the negligible skin contact from these types of products, there is no justification for a restriction of the concentration of this fragrance ingredient in the finished products	Sensitization
28.	Treemoss extracts	Restricted/ Specification	Due to the negligible skin contact from these types of products there is no justification for a restriction of the concentration of this fragrance ingredient in the finished product	Sensitization
29.	Verbena oil	Prohibited	Leave on products: n/a Non skin contact products: n/a	Sensitization, photoxicity
30.	Ylang-ylang extracts	Restricted	Due to the negligible skin contact from these types of products, there is no justification for a restriction of the concentration of this fragrance ingredient in the finished product	Sensitization

7.3 CASE STUDIES OF MULTINATIONAL COMPANIES

The global cosmetics industry is extremely regulated to cinch consumer safety, particularly when it comes to the use of essential oils and fragrances. Regulatory bodies such as the FDA in the United States, the European Medicines Agency (EMA) in the EU, and others across the globe have exacting guidelines that multinational companies must follow. This explores the foundational regulatory frameworks that steer the formulation, labelling, and marketing of cosmetics containing essential oils and fragrances. Understanding these practices is pivotal for companies to avoid legal pitfalls and ensure product compliance (European Medicines Agency – EMA | European Union, n.d.).

7.3.1 L'Oréal's compliance with EU cosmetic regulations

L'Oréal, a leader in the cosmetics industry, has established a strong compliance framework to hew to the EU's stringent cosmetic regulations, particularly the EU Cosmetics Regulation (EC) No. 1223/2009. The exploration delves into L'Oréal voyages, the complex landscape of ingredient safety assessments, product labelling, and marketing claims for its products containing essential oils and fragrances. The focus is on L'Oréal's strategies for maintaining compliance across diverse markets in the EU while nurturing innovation in product development (L'Oréal, World Leader in Beauty : Makeup, Cosmetics, Haircare, Perfume, n.d.). L'Oréal ensures food-grade or beauty-grade packaging quality, complies with European regulations, and ceased using PVC in 2018. By 2030, it aims for 100% traceable, sustainable bio-sourced packaging. In 2022, 99.6% of its wood-based packaging was ascribable, with 89.2% certified by FSC or PEFC. L'Oréal promotes FSC standards globally. The compliance strategy of L'Oréal responded by redrafting some of its products to decrease the concentration of allergens below the threshold level and contemporize its product labelling across the EU market. They also hoarded in educating their research and development teams about the elaborating regulatory landscape. L'Oréal's proactive avenue ensured continued compliance with the EU regulations, and eschewed potential product reminisce or market withdrawals. It demonstrates the importance of abiding ahead of regulatory alteration in a highly regulated market like the EU (4.3 Policies, Performance Indicators and Results - Page 213 | L'Oréal - 2023 Universal Registration Document, n.d.).

7.3.2 Estée Lauder's avenue to FDA regulations in the United States

Estée Lauder wields within a complex regulatory environment in the United States, governed primarily by the FDA. The company ensures compliance with regulations that govern the use of essential oils and fragrances in cosmetics, including Estée Lauder's ingredient safety testing protocols, adherence to the VCRP, and their strategies for managing fragrance allergens. It also studies how Estée Lauder stays ahead of regulatory changes and maintains consumer trust in a highly competitive market (Voluntary Cosmetic Registration Program | FDA, n.d.). The compliance strategy of Estée Lauder hoarded local research and development and partnered with Chinese regulatory experts to ensure its products confronted the safety standards required by the NMPA. The company also worked meticulously with Chinese authorities to navigate the regulatory approbation process, including conducting additional testing. Estée Lauder's steadfastness to understanding and meeting local regulatory requirements endorsed it to successfully enter and expand in the Chinese market. It demonstrates the importance of localized proficiency and collaboration with regulatory bodies in conquering compliance challenges.

7.3.3 Shiseido's compliance with Japanese and Chinese regulations

Shiseido, a major player in the cosmetics industry, faces outlandish regulatory challenges in Asia, particularly in Japan and China. This research explores that Shiseido ensures its products containing essential oils and fragrances comply with Japan's Pharmaceuticals and Medical Devices Act (PMD Act) and China's National Medical Products Administration (NMPA) regulations. The study focuses on Shiseido's strategies for ingredient safety evaluation, product registration, and navigating the complexities of cross-border regulatory compliance in two of the largest cosmetics markets in the world. Shiseido entered the Chinese market in the 1980s, initially succeeding with its AUPRES brand, developed exclusively for China. To maintain its high-end image, Shiseido limited retail expansion, differentiating AUPRES from its luxury SHISEIDO brand (Shiseido Company, n.d.). Positioned between expensive imports and cheap local products, AUPRES targeted the middle segment. With increased competition from brands like P&G and Estée Lauder, Shiseido expanded its product line and utilized a unique channel strategy involving speciality shops, a model imported from Japan. Recently, Shiseido also began selling through drugstores and online platforms. This investigation heightens questions about the merits and challenges of Shiseido's expansion in China compared to Europe and the United States, the effectiveness of its marketing strategies, and its long-term differentiation strategy as a "global player representing Asia" amidst evolving global market trends (Welcome to Ministry of Health, Labour and Welfare, n.d.).

7.3.4 Unilever's sustainable netting and compliance with global fragrance regulations

Unilever has made sustainability a core part of its business strategy, particularly in sourcing essential oils for its fragrance products. This scrutiny demonstrates Unilever integrates sustainability into its compliance practices across global markets. The exploration focuses on Unilever's adherence to international fragrance regulations, including the IFRA standards, and how the company ensures that its sourcing practices for essential oils align with regulatory requirements. The study also investigates how Unilever balances sustainability, regulatory compliance, and consumer expectations in the global cosmetics market. According to its compliance strategy, Unilever developed a global compliance framework that endorsed it to adapt its products and marketing strategies to meet the regulatory requirements of each market. This included various formulations and labelling for different regions, as well as peculiar training for local teams on regulatory compliance. Unilever's ability to supervise a complex regulatory landscape across multiple markets permitted the company to maintain a strong global presence without facing significant compliance-related convulsion. This exploration accentuates the need for a flexible and localized avenue to regulatory compliance in multinational operations (Unilever Global: Making Sustainable Living Commonplace | Unilever, n.d.).

7.3.5 Johnson & Johnson's acclimatization to regulatory changes in the European market

Johnson & Johnson has had to adapt quickly to regulatory changes in the European market, particularly concerning essential oils and fragrances. This exploration proves the company has navigated the updates to the EU Cosmetics Regulation, including changes in ingredient restrictions and labelling requirements. The assessment emphasizes Johnson & Johnson's proactive approach to product reformulation, compliance monitoring, and its strategies for regulating regulatory risks. Accoding to its compliance strategy, Johnson & Johnson began a voluntary recall of the grieved product batches and regulated a thorough investigation into the induce of allergic reactions. The company also reviewed its conceptualization process and strengthened its allergen testing protocols. It is immersed in transparency with both consumers and regulators about the steps being taken to prevent future incidents. Johnson & Johnson's hasty and transparent response helped to allay the negative impact of the situation, preserving consumer trust and being docile with regulatory requirements. The critical importance of effective pitfall management and communication in handling regulatory compliance problems (Our Company, n.d.).

7.3.6 Chanel's compliance with international standards for fragrance allergens

Chanel, known for its luxury fragrances, must navigate complex international standards for fragrance allergens. This exploration ensures Chanel's compliance with regulations such as the EU's Cosmetic Products

Regulation and the IFRA standards (IFRA Newsroom, n.d.). The focus is on Chanel's methods for identifying and managing allergens in their products, the challenges of maintaining product appropriateness across different markets, and the company's strategies for labelling and consumer safety. The dissection explores how Chanel balances the artistic veracity of its fragrances with the need for regulatory compliance (Chanel Official Website: Fashion, Fragrance, Beauty, Watches, Fine Jewelry | Chanel, n.d.).

7.3.7 Avon's regulatory compliance strategy for essential oils in Latin America

Avon, a major player in the direct-selling cosmetics industry, faces unique regulatory challenges in Latin America. This exploration navigates the complex regulatory environments of countries like Brazil, Mexico, and Argentina, where the use of essential oils in cosmetics is meticulously monitored. This scrutiny explores Avon's approach to ingredient approval, product safety assessments, and compliance with local labelling requirements. Furthermore, it highlights Avon leverages its regional proficiency to maintain compliance and build consumer trust in markets with diverse regulatory landscapes. Avon's use of regional expertise and targeted compliance strategies enables it to build consumer trust and product safety (PAHO/WHO | Pan American Health Organization, n.d.).

7.3.8 The Body Shop's ethical sourcing and compliance with global cosmetic regulations

The Body Shop, known for its allegiance to ethical sourcing, faces the challenge of ensuring that its use of essential oils in cosmetics complies with global regulations. This appraisal demonstrates whether the Body Shop aligns its ethical sourcing practices with compliance requirements in markets such as the EU, the United States, and Asia. It includes the Body Shop's strategies for ensuring ingredients are verifiable, conducting safety assessments, and adhering to labelling regulations. This assessment proves that the company maintains its ethical stance while navigating the complexities of international regulatory compliance (Cruelty-Free & Beauty Products | The Body Shop, n.d.).

7.3.9 Procter & Gamble's global strategy for regulatory compliance in surfacing markets

Procter & Gamble (P&G) has a robust strategy for regulatory compliance, particularly in emerging markets. This dissection examines P&G's approach to meeting the diverse regulatory requirements of countries in Latin America, Africa, and Southeast Asia. The deliberation covers P&G's

managing the challenges of ingredient approval, product labelling, and consumer safety in regions where regulatory frameworks are still evolving. This also emphasizes P&G's proactive engagement with local regulatory bodies and its efforts to educate consumers about the safety of its products (Procter & Gamble, n.d.).

7.4 CONCLUSION

The global regulatory landscape for essential oils and fragrances in cosmetics is complex and varies by region. Multinational companies must stay informed about local regulations, conduct rigorous safety assessments, implement good manufacturing practices (GMP), ensure accurate labelling, and maintain thorough documentation to ensure compliance. By adopting these strategies, companies can navigate regulatory challenges and ensure the safety and quality of their cosmetic products worldwide. Regulatory framework particularly in the United States and Europe and stringent with regulations like REACH, FDA 21 CFR, and California Proposition 65. Overall essential oils continue to thrive in the cosmetics industry; sustainability practices and regulation compliance are critical for long-term success and environmental preservation.

REFERENCES

4.3 Policies, performance indicators and results - Page 213 | L'Oréal - 2023 universal registration document. (n.d.). Retrieved September 7, 2024, from https://www.loreal-finance.com/eng/2023-universal-registration-document/en/article/213/

Apply for FSSAI Food Licenses Online | Food Licence Apply. (n.d.). Retrieved September 7, 2024, from https://foodlicenceapply.com/?gad_source=1&gclid=Cj0KCQjw8–2BhCHARIsAF_w1gw72isH_L8tkwuyjiuvBL08ll_r7ZUwrSdIaEPSmfneCJd_Yct_4MUaAm5sEALw_wcB

Babita, S., Sellam, P., Jayoti, M., & Puja, R. (2014). Floral essential oils: Importance and uses for mankind. *Hort Flora Research Spectrum*, *3*, 7–13.

Bakkali, F., Averbeck, S., Averbeck, D., & Idaomar, M. (2008). Biological effects of essential oils–a review. *Food and Chemical Toxicology*, *46*(2), 446–475.

Barbieri, C., Borsotto, P., & others. (2018). Essential oils: Market and legislation. *Potential of Essential Oils*, 107–127.

Ben Hsouna, A., Ben Halima, N., Smaoui, S., & Hamdi, N. (2017). Citrus lemon essential oil: Chemical composition, antioxidant and antimicrobial activities with its preservative effect against *Listeria monocytogenes* inoculated in minced beef meat. *Lipids in Health and Disease*, *16*, 1–11.

Burger, P., Plainfossé, H., Brochet, X., Chemat, F., & Fernandez, X. (2019). Extraction of natural fragrance ingredients: History overview and future trends. *Chemistry\& Biodiversity*, *16*(10), e1900424.

CDSCO. (n.d.). Retrieved September 7, 2024, from https://www.cdscoonline.gov.in/CDSCO/homepage

Chanel Official Website: Fashion, Fragrance, Beauty, Watches, Fine Jewelry | Chanel. (n.d.). Retrieved September 7, 2024, from https://www.chanel.com/in/

China's New Cosmetics Regulation (CSAR). (n.d.). Retrieved September 7, 2024, from https://www.intertek.com/assuris/cosmetics/regulatory/china-cosmetic-regulation-csar/

Coppen, J. J. W. (1994). *Flavours and fragrances of plant origin: A review of the production, markets and development potential of selected essential oils and resins and their plant sources. Study undertaken for the Food and Agriculture Organization of the United Nations.*

Cosmetics Regulation and REACH-questions and answers on animal testing What does REACH say about animal testing? (n.d.).

Cruelty-Free & Beauty Products | The Body Shop. (n.d.). Retrieved September 7, 2024, from https://www.thebodyshop.in/

CSAR: China's New Cosmetic Regulation and What It Means. (n.d.). Retrieved September 7, 2024, from https://www.linkedin.com/pulse/csar-chinas-new-cosmetic-regulation-what-means-coptis

Do I Need a License to Sell Essential Oils in India - Oil Therapy Central. (n.d.). Retrieved September 7, 2024, from https://oiltherapycentral.com/blogs/essential-oils/license-sell-essential-oils-india/

eCFR :: 21 CFR Part 700 Subpart A – General Provisions. (n.d.). Retrieved September 7, 2024, from https://www.ecfr.gov/current/title-21/chapter-I/subchapter-G/part-700/subpart-A

Entering the European market for essential oils | CBI. (n.d.). Retrieved September 7, 2024, from https://www.cbi.eu/market-information/natural-ingredients-cosmetics/essential-oils/market-entry

Essential Oil Regulations in the EU: A Guide. (n.d.). Retrieved September 7, 2024, from https://www.linkedin.com/pulse/essential-oil-regulations-eu-guide-intended-uses-regulatory-uu18f

Essential Oil Regulations in the United States: An Overview. (n.d.). Retrieved September 7, 2024, from https://www.compliancegate.com/essential-oil-regulations-united-states/

Essential Oils and Resinoids: Top Importers Worldwide 2021 | Statista. (n.d.). Retrieved September 7, 2024, from https://www.statista.com/statistics/973314/global-top-importers-of-essential-oils-and-resinoids/

Essential Oils Market Size, Share & Growth Report, 2030. (n.d.). Retrieved September 7, 2024, from https://www.grandviewresearch.com/industry-analysis/essential-oils-market

European Medicines Agency – EMA | European Union. (n.d.). Retrieved September 7, 2024, from https://european-union.europa.eu/institutions-law-budget/institutions-and-bodies/search-all-eu-institutions-and-bodies/european-medicines-agency-ema_en

Fair Packaging and Labeling Act: Regulations Under Section 4 of the Fair Packaging and Labeling Act | Federal Trade Commission. (n.d.). Retrieved September 7, 2024, from https://www.ftc.gov/legal-library/browse/rules/fair-packaging-labeling-act-regulations-under-section-4-fair-packaging-labeling-act

Federal Hazardous Substances Act (FHSA) Requirements | CPSC.gov. (n.d.). Retrieved September 7, 2024, from https://www.cpsc.gov/Business–Manufacturing/Business-Education/Business-Guidance/FHSA-Requirements

FFDC Kannauj | Kannauj Attar. (n.d.). Retrieved September 7, 2024, from https://kannaujattar.com/ffdc-kannauj/

Fonseca-Santos, B., Corrêa, M. A., & Chorilli, M. (2015). Sustainability, natural and organic cosmetics: Consumer, products, efficacy, toxicological and regulatory considerations. *Brazilian Journal of Pharmaceutical Sciences*, *51*(1), 17–26.

Guirguis, K., Gershunov, A., Tardy, A., & Basu, R. (2014). The impact of recent heat waves on human health in California. *Journal of Applied Meteorology and Climatology*, *53*(1), 3–19. https://doi.org/10.1175/JAMC-D-13-0130.1

Ibrahim, F. M., Mohammed, R. S., Abdelsalam, E., Ashour, W. E.-S., Magalhães, D., Pintado, M., & El Habbasha, E. S. (2024). Egyptian citrus essential oils recovered from lemon, Orange, and mandarin peels: Phytochemical and biological value. *Horticulturae*, *10*(2), 180.

IFRA Newsroom. (n.d.). Retrieved September 7, 2024, from https://ifrafragrance.org/news/newsroom

Inoue, M., & Hayashi, S. (2023). Opportunities and challenges for medicinal and aromatic plants in Japan. *Plants as Medicine and aromatics*, 173–186.

Introduction. (n.d.). Retrieved September 7, 2024, from https://ifrafragrance.org/policy/introduction

ISO/TC 54 - Essential oils. (n.d.). Retrieved September 7, 2024, from https://www.iso.org/committee/48956/x/catalogue/

Japan Cosmetic Regulation | ChemLinked. (n.d.). Retrieved September 7, 2024, from https://cosmetic.chemlinked.com/cosmepedia/japan-cosmetic-regulation

Japan External Trade Organization (JETRO) Development Cooperation Division Trade and Economic Cooperation Department. (2011).

L'Oréal, world leader in beauty: makeup, cosmetics, haircare, perfume. (n.d.). Retrieved September 7, 2024, from https://www.loreal.com/en/

Our Company. (n.d.). Retrieved September 7, 2024, from https://www.jnj.com/our-company

PAHO/WHO | Pan American Health Organization. (n.d.). Retrieved September 7, 2024, from https://www.paho.org/en

Procter & Gamble. (n.d.). Retrieved September 7, 2024, from https://us.pg.com/

Rayasam, S. D. G., Koman, P. D., Axelrad, D. A., Woodruff, T. J., & Chartres, N. (2022). Toxic Substances Control Act (TSCA) implementation: How the amended law has failed to protect vulnerable populations from toxic chemicals in the United States. *Environmental Science\& Technology*, *56*(17), 11969–11982.

Raymakers, C. (2006). CITES, the Convention on International Trade in Endangered Species of Wild Fauna and Flora: Its role in the conservation of *Acipenseriformes*. *Journal of Applied Ichthyology*, *22*.

Sharifi-Rad, J., Sureda, A., Tenore, G. C., Daglia, M., Sharifi-Rad, M., Valussi, M., Tundis, R., Sharifi-Rad, M., Loizzo, M. R., Ademiluyi, A. O., & others. (2017). Biological activities of essential oils: From plant chemoecology to traditional healing systems. *Molecules*, *22*(1), 70.

Sharmeen, J. B., Mahomoodally, F. M., Zengin, G., & Maggi, F. (2021). Essential oils as natural sources of fragrance compounds for cosmetics and cosmeceuticals. *Molecules*, *26*(3), 666.

Shiseido Company. (n.d.). Retrieved September 7, 2024, from https://corp.shiseido.com/en/

Tripathi, H., Suresh, R., Kumar, S., & Khan, F. (2016). International trade in essential oils and resinoids: A case study of past 18 years. *Journal of Medicinal and Aromatic Plant Sciences*, *38*(1–4), 1–28.

Turek, C., & Stintzing, F. C. (2013). Stability of essential oils: A review. *Comprehensive Reviews in Food Science and Food Safety*, *12*(1), 40–53.

Unilever Global: Making sustainable living commonplace | Unilever. (n.d.). Retrieved September 7, 2024, from https://www.unilever.com/

Voluntary Cosmetic Registration Program | FDA. (n.d.). Retrieved September 7, 2024, from https://www.fda.gov/cosmetics/voluntary-cosmetic-registration-program

Welcome to Ministry of Health, Labour and Welfare. (n.d.). Retrieved September 7, 2024, from https://www.mhlw.go.jp/english/

Winkelman, W. J. (2018). Aromatherapy, botanicals, and essential oils in acne. *Clinics in Dermatology*, *36*(3), 299–305.

Wongsukkasem, N., Soynark, O., Suthakitmanus, M., Chongdiloet, E., Chairattanapituk, C., Vattanikitsiri, P., Hongratanaworakit, T., & Tadtong, S. (2018). Antiacne-causing bacteria, antioxidant, anti-tyrosinase, anti-elastase and anti-collagenase activities of blend essential oil comprising rose, bergamot and patchouli oils. *Natural Product Communications*, *13*(5), 1934578X1801300529.

Chapter 8

Innovations and regulatory responses to the use of essential oils and fragrances in cosmetic products

Balancing consumer safety and industry advancement

Saniya Shakir, Samreen Abbas, Saumya Gupta, Neha Verma, and Bhartendu Nath Mishra
Department of Biotechnology, Institute of Engineering and Technology, Lucknow, India

Shiv Poojan
bioTOX Solution, NNF, IET, Lucknow, India

8.1 INTRODUCTION

The cosmetic industry has witnessed a growing interest in the use of natural ingredients, particularly essential oils (EOs) and plant-based fragrances, to meet the consumers increasing demand for more gentler formulations which are sustainable and eco-friendly (Berger, 2007). The utilization of natural ingredients in cosmetics, such as EOs and plant-based fragrances, is not a new concept; there are different cosmetic products containing these already available in the market, as shown in Figure 8.1, but their utilization has increased significantly in recent years. These natural ingredients offer potential benefits, such as improved skin hydration, antimicrobial properties, and enhanced sensorial experiences (Aburjai & Natsheh, 2003).

Cosmetology, the study and application of beauty treatments, has evolved to encompass not only the enhancement of physical appearance but also a greater emphasis on consumer health and environmental responsibility. The move toward natural cosmetics has challenged ingredient suppliers to adapt and innovate. Traditionally, cosmetics were expected to cover skin imperfections, add color, or impart a pleasant scent. However, in recent years, there has been a shift toward developing cosmetics that claim to incorporate new technology and promise to do something good for the user's body. This movement is continuing to strengthen as marketers respond to consumer demand for products containing more botanicals and natural ingredients. Keeping in line with a "greener" public, these products are designed to be more in

 DOI: 10.1201/9781003558477-8

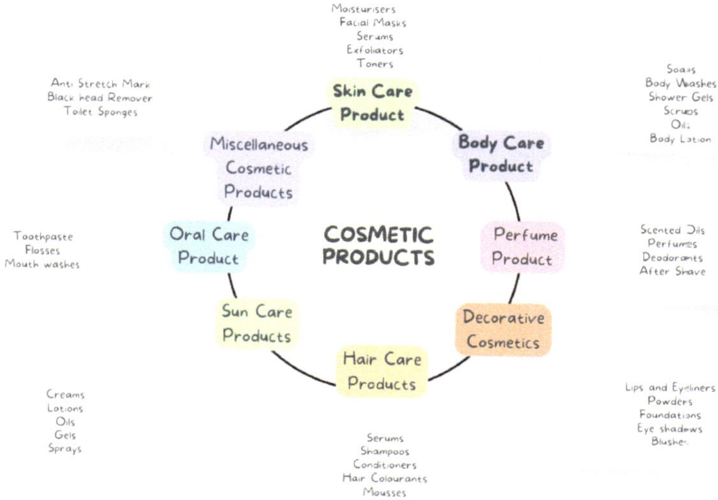

Figure 8.1 Different types of cosmetic products containing essential oils and fragrances.

harmony with nature, featuring recyclable packaging and claims of not having been tested on animals (Hosoi et al., 2017).

As the consumer demand for natural and sustainable cosmetics grows, the industry has sought to innovate and expand the use of EOs and plant-based fragrances. These natural ingredients are not only aesthetically pleasing but also offer potential therapeutic benefits, such as promoting skin health or eliciting positive emotional responses. Manufacturers are working to incorporate EOs and botanicals into a wide range of cosmetic products, from moisturizers and serums to makeup and haircare items. This trend reflects a broader shift toward more holistic approaches to beauty, where consumers seek products that align with their values of environmental stewardship and personal wellbeing (Carvalho et al., 2016). Moreover, the increasing consumer awareness of the benefits of organic and natural ingredients has led to a surge in research and development within the cosmetic industry, fostering innovation in formulations that not only enhance beauty but also contribute positively to skin health and overall wellbeing (Duran et al., 2014). Furthermore, EOs have emerged as key components in these formulations, recognized not only for their aromatic properties but also for their bioactive compounds that can provide various skin benefits, including anti-inflammatory and antioxidant effects, thereby aligning with the growing demand for multifunctional cosmetic products.

Beyond traditional cosmetic applications, the industry has also witnessed the rise of cosmeceuticals and nutricosmetics, which represent a convergence of cosmetics and pharmaceuticals. These products leverage active ingredients like EOs to deliver enhanced benefits, such as skin repair, improved elasticity, and anti-aging effects, effectively blurring the lines between beauty and therapeutic treatment options offered to consumers. This innovative approach not only caters to the cosmetic needs of consumers but also addresses their demand for safe and effective solutions, as research continues to highlight the potential of plant-derived bioactive compounds in promoting skin health, beauty, and overall wellbeing. The expansion of cosmeceuticals reflects a growing trend toward products that are not only aesthetically beneficial but also support skin health and rejuvenation through scientifically validated ingredients, thus responding to an increasingly informed consumer base seeking both performance and efficacy in their beauty regimen (López-Hortas et al., 2021). As the market for cosmeceuticals continues to expand, understanding the role of phytochemicals and EOs in formulation will be crucial for developing innovative products that meet consumer demands and regulatory standards, thereby ensuring their effectiveness and safety in skincare applications.

This chapter explores the innovations in the use of EOs and fragrances in cosmetic products, the regulatory responses to address the associated risks, and the challenges of maintaining a balance between consumer protection and industry progress. It examines the growing consumer demand for more sustainable and natural cosmetic formulations, and how the cosmetic industry has responded by incorporating EOs and fragrances into their products. Furthermore, it delves into the regulatory frameworks that have been established to evaluate the safety and efficacy of these cosmeceuticals, highlighting the ongoing efforts to strike a balance between fostering industry innovation and safeguarding consumers.

8.1.1 Background and significance

The background and significance of EOs and fragrances in the cosmetic industry cannot be overstated, as these natural extracts play a vital role in product formulation and consumer appeal. EOs and their components are not only appreciated for their aromatic properties but are also recognized for their wide-ranging functional benefits, such as analgesic, antiseptic, antimicrobial, and anti-inflammatory properties, making them ideal candidates for a variety of cosmetic applications. These beneficial properties stem from the complex chemical compositions of EOs, which can include terpenes, alcohols, esters, and other bioactive compounds.

As the demand for botanical extracts in cosmetic products grows, the industry increasingly seeks to exploit the biological properties of these ingredients to enhance human health, beauty, and wellness. The versatility of EOs has led to their incorporation in haircare and skincare products, where they serve multiple roles, including acting as natural preservatives, promoting hair growth, and improving skin health. For instance, peppermint oil has been shown to effectively stimulate hair growth when applied topically at low concentrations, comparable to traditional treatments like minoxidil.

Additionally, certain components of EOs, such as menthol and camphor, enhance blood circulation in hair follicles, which is crucial for maintaining healthy hair growth and preventing loss (Guzmán & Lucia, 2021).

However, the utilization of EOs in cosmetics also necessitates caution. The International Fragrance Association (IFRA) has established maximum concentration levels for these oils to mitigate potential allergy risks associated with their use. As the field continues to evolve, it becomes essential to employ advanced analytical techniques to ensure the purity and safety of EOs, addressing concerns related to adulteration and allergenic potential. Thus, the background and significance of EOs in cosmetics reflect a delicate balance between leveraging their beneficial properties and ensuring consumer safety and satisfaction (Galea et al., 2023).

8.2 INNOVATIONS IN ESSENTIAL OILS AND FRAGRANCES

The cosmetic industry has long been driven by consumer demand for natural and eco-friendly products, leading to a growing interest in the use of EOs and plant-based fragrances. However, the inherent limitations of these natural ingredients, such as instability and low skin penetration, have prompted the industry to explore new technologies to address these challenges (Sharmeen et al., 2021). Recent technological advancements have dramatically transformed the formulation of cosmetic products, particularly through innovations in customization and efficiency. These advancements in the formulation of EOs and fragrances for cosmetic applications have ushered in a new era for the beauty and personal care industry. As the cosmetic industry continues to evolve, these technological advancements serve as key drivers for growth, enabling the rapid introduction of innovative products while ensuring compliance with safety regulations (Baki & Peeters, 2019). There are many cosmetic products available in the market which use innovative techniques; some of those products are listed in Table 8.1. Ultimately, the interplay of advanced technologies in cosmetic formulation not only enhances product efficacy but also enriches the consumer experience in a highly competitive market.

Innovations in the cosmetic industry have significantly transformed the sensory experience associated with beauty products, making every day routines more pleasurable and effective. The concept of sensoriality goes beyond just the functional benefits of a product; it creates a sensory journey in which various elements combine to enhance user satisfaction. For instance, the development of unique textures plays a crucial role in this experience, as pleasant textures can elevate the feeling of indulgence during application, turning routine into a sensory delight. Moreover, fragrances are essential in this transformation; the uplifting scent of floral notes not only enhances enjoyment but also contributes to the overall sensory experience, making self-care feel more luxurious. Importantly, these innovations also focus on improving efficacy, as

Table 8.1 List of cosmetic products available in the market which are manufactured using emerging trends and technologies

S. No.	*Product*	*Brand*	*Material*	*Advantages*	*Disadvantages*	*Size*	*Reference*
1.	Concealer	Clinique, Clarins, L'Oréal, Revlon, The Body Shop, Max Factor, Lancôme Paris and By Terry	Silica solid lipid nanoparticles	High coverage Appearance Adherence to the skin	Toxicity DNA damage Oxidative stress	Less than 100 nm	Gupta et al. (2022); *Nanoparticles Found in 10 Top Brand Cosmetics – Emerging Technology* (n.d.)
2.	Foundation	Clinique, Clarins, L'Oréal, Revlon, The Body Shop, Max Factor, Lancôme Paris, By Terry, Yves Saint Laurent and Christian Dior	Titanium oxide and zinc oxide	Texture and transparency Appearance coverage Dermal penetration	Increases concentration of active ingredient in blood causes toxicity Environmental concerns Teratogenic in nature	Less than 100 nm	Gupta et al. (2022); *Nanoparticles Found in 10 Top Brand Cosmetics – Emerging Technology* (n.d.)
3.	Sunscreen	La Roche Posay, L'Oréal, Biotherm, Vichy	TiO_2, ZnO	UV blockers No white cast Shelf life of product	Toxicity due to high surface area to volume ration Increased penetration of nanoparticles affects the cells of the human body		Santos et al. (2022) Yadwade et al. (2021)
4.	Anti-aging cream	Christian Dior	Liposome	Non- toxic Biodegradable Flexible Encapsulate active component for easy and safe delivery	Oxidation of liposome causes deterioration in quality of liposomes		Fytianos et al. (2020) Yadwade et al. (2021)

5.	Lipstick	Sugar cosmetics	Silica	Enhances texture and aesthetics Provides an SPF boost Minimizes pores Absorbs oils	It can cause skin allergies like itching, burning, redness, etc.	Fytianos et al. (2020) Yadwade et al. (2021)
6.	Perfume	Chanel Allur	Solid lipid nanoparticles	Better stability Bioavailability Augmented permeability Enhanced solubility	Limited loading capacity Poor stability under acidic condition	(Chadha (2021) Hooda & Sradhanjali (2017)
7.	Anti-aging cream	L'Oréal, Geltor, Modern Meadow, Jellatech	Synthetic biology (collagen)	Improve skin elasticity Provide hydration to the skin Repairs and regenerates skin	Limited penetration Effectiveness of collagen depends on its form	Jadach et al. (2024); *Synthetic Biology in Everyday Life: From Lab to the Beauty Counter - SynBioBeta* (n.d.)
8.	Skin brightening cream	Amyris, Givaudan, Biossance	Synthetic biology (squalan)	Natural antioxidant and antitumor properties Excellent emollient Boost hydration	May cause allergies like itching, redness, and swelling	*Synthetic Biology in Everyday Life: From Lab to the Beauty Counter - SynBioBeta* (n.d.); Yadwade et al. (2021)
9.	Haircare and skincare	K18 Hair (take over by Unilever), Ourself	Synthetic biology (bioactive peptide)	Wound healing properties Collagen stimulating Repair Antimicrobial properties Enhanced skin permeability	Poor solubility at high concentration Issue in maintaining the structural stability Heavily influenced by various variables like temperature and Ph	Ngoc et al. (2023) *Synthetic Biology in Everyday Life: From Lab to the Beauty Counter - SynBioBeta* (n.d.)

(*Continued*)

Table 8.1 (Continued) List of cosmetic products available in the market which are manufactured using emerging trends and technologies

S. No.	*Product*	*Brand*	*Material*	*Advantages*	*Disadvantages*	*Size*	*Reference*
10.	Exfoliating mask	Pilgrim, Plix, Minimalist, Dermalogica	Alpha-hydroxy acids, beta-hydroxyl acid	Sensitive to UV light Inflammatory properties Penetrates the skin Moisturizing property	High concentration AHA will disrupt the cohesion of corneocytes of the skin barrier and causes skin irritation		Tang & Yang (2018); *What Are Chemical Exfoliants for Skin? | CeraVe Australia* (n.d.) Yadwade et al. (2021)
11.	Skin brightening cream	Primex (Chito care beauty), Madara, Naturopathica	Chitosan	Anti-microbial Moisture retentive Film forming Collagen maintenance	Interaction with cells may be altered due to the presence of charged particles in chitosan		EWG (n.d.) Guzmán et al. (2022)
12.	Moisturizer	Mamaearth, Chanel, Dot & Key, Derma Co	Hyaluronic acid	It makes skin flexible Stretch skin Wound healing and reduces scarring	May cause allergic responses like pain, swelling, and redness.		*Amazon.in: Hyaluronic Acid Moisturizer* (n.d.); *Hyaluronic Acid: Uses, Side Effects, and Risks* (n.d.)
13.	Perfume	Orebella, Phlur, The 7 Virtues, Sniff, Prada, DedCool, Ceremonia, Future Society	Essential oils	Relieve stress Freshen the skin	Burning and irritation Asthma attack Headache		*11 Gorgeous Perfumes Made from Essential Oils* (n.d.); (What Are Essential Oils? Uses and Side Effects (n.d.)
14.	Anti-acne product	Dot & Key, The pink foundry, Hyphen, MFine	Salicylic acid	Rejuvenate and exfoliate the skin Anti-fungal and anti- infection	Irritation Dryness Increases warmth of the skin		*Amazon.in: Hyaluronic Acid Moisturizer* (n.d.); What Are Essential Oils? Uses and Side Effects (n.d.)

the integration of multifunctional ingredients, like humectants and conditioning polymers, enriches both texture and effectiveness, ensuring that products perform well while providing a pleasurable experience. The strategic use of texture modifiers, such as glycerin and silicones, further enhances the tactile sensation of products, ensuring a smooth and velvety finish that users find appealing. Overall, the emphasis on sensory experiences is shaping the future of the beauty and personal care industry, driving companies to innovate and create products that delight the senses while delivering effective results (Stone et al., 2020).

8.2.1 Impacts of emerging trends and technologies

8.2.1.1 Nanotechnology

Nanotechnology has the potential to enhance the efficacy and stability of natural cosmetic ingredients. It has not only expanded the categories of cosmetic products but also enhanced their therapeutic benefits. The use of nanosized ingredients has yielded formulations known as nanocosmeceuticals, which utilize innovative delivery mechanisms that improve the absorption and effectiveness of active ingredients (Gupta et al., 2022). Nanoparticles can be used to encapsulate and deliver EOs and fragrances more effectively, improving their absorption and prolonging their effects on the skin. Nanoemulsions, liposomes, and solid lipid nanoparticles are some of the nanocarrier systems that have been investigated for this purpose, offering improved stability, targeted delivery, and controlled release of active ingredients (Kan et al., 2024).

The incorporation of nanotechnology in cosmeceuticals has been driven by the belief that smaller particles are more readily absorbed into the skin, allowing for more efficient repair and protection. Some of the key advantages of using nanoparticles in cosmetic products include improved ingredient stability, enhanced UV protection, and a more aesthetically pleasing appearance (Fytianos et al., 2020). The properties of cosmetic ingredients are notably altered at the nanoscale, leading to better solubility, stability, and chemical reactivity, all of which contribute to the enhanced performance of these products and transform consumer expectations and experiences. The incorporation of EOs into products like anti-aging serums and facial oils has gained traction, with oils such as rosehip seed and chamomile showing significant improvements in skin texture and hydration (Dubey et al., 2022). The trend also extends to lipstick formulations, where EOs are now being used to impart specific flavors and serve as dispersants for colorants, enhancing both the sensory experience and functional properties of these products.

However, the use of nanoparticles in cosmetics has also raised concerns about their potential toxicity and impact on human health. Nanotoxicological research has indicated that there are possibilities of nanoparticles to penetrate through the skin and cause health hazards, leading to increased regulatory scrutiny and consumer awareness (Kaul et al., 2018). To address these concerns, the cosmetic industry is exploring the use of natural nanomaterials, such as those derived from plants and marine sources, which may offer a more sustainable and safer alternative. The application of nanotechnology to natural cosmeceuticals has the potential to improve the stability and skin penetration of these ingredients, while mitigating the risks associated with synthetic nanoparticles.

In conclusion, the impact of new technologies, particularly nanotechnology, on the use of EOs and fragrances in cosmetic products is a complex and evolving landscape. While these technologies offer significant benefits in terms of efficacy and stability, the potential risks associated with nanoparticles have led to increased regulatory and consumer scrutiny, driving the industry to explore more sustainable and safe alternatives (Chauhan & Chauhan, 2023).

8.2.1.2 Synthetic biology

The use of synthetic biology has also emerged as a promising approach for enhancing the production and performance of natural cosmetic ingredients. Through the use of advanced techniques like microbial engineering and metabolic engineering, synthetic biology can facilitate the creation of plant-based compounds, such as EOs and fragrances, with improved stability, consistency, and cost-effectiveness. By engineering microorganisms to produce these natural-like compounds, the cosmetic industry can overcome the challenges associated with traditional extraction methods, which can be heavily influenced by factors like climate, geography, and seasonal variations (Gupta et al., 2019). Furthermore, synthetic biology provides the opportunity to design and develop unique molecular structures that mimic the properties of natural ingredients but possess enhanced skin penetration, increased efficacy, or other desirable properties that may not be readily available in naturally occurring ingredients. This level of customization and optimization enables the formulation of cosmetic products that are more targeted, effective, and sustainable than those relying solely on natural resources. Overall, the application of synthetic biology represents a promising pathway for the cosmetic industry to meet consumer demand for natural-inspired products while overcoming the inherent limitations of nature (Gomes et al., 2020).

While synthetic biology offers the potential to design and develop unique molecular structures with enhanced properties, this approach also raises concerns about the authenticity and transparency of natural-inspired cosmetic products. The ability to engineer microorganisms to produce plant-based compounds that mimic natural ingredients may lead to disconnect between consumer expectations and the true origins of the ingredients used. This could undermine the trust and integrity of the natural and eco-friendly ethos that many consumers seek in their cosmetic purchases. Additionally, the long-term effects of these synthetic biology-derived compounds on human health and the environment are not yet fully understood, warranting cautious consideration of their widespread adoption in the cosmetic industry. Ultimately, any benefits of increased customization and optimization must be carefully weighed against the potential risks and the need to maintain consumer confidence in the authenticity and sustainability of natural cosmetic products (Butschke et al., 2016).

It is crucial for the cosmetic industry to proactively address the concerns surrounding the use of emerging technologies, such as nanotechnology and synthetic biology, in the development of natural-inspired cosmetic products. This should be done through a multi-pronged approach that prioritizes transparency, consumer education, and rigorous safety testing. Clear and comprehensive labeling of product ingredients, including whether any synthetic or engineered components are present, will help maintain consumer trust and confidence in the authenticity of natural cosmetic formulations. Furthermore, the industry should invest in educational initiatives that inform consumers about the benefits and potential risks associated with these innovative technologies, empowering them to make informed purchasing decisions. Lastly, the industry must undertake rigorous safety testing to thoroughly evaluate the long-term impacts of these novel ingredients on human health and the environment, ensuring their responsible and ethical use in cosmetic products. By adopting these measures, the industry can navigate the complexities of incorporating new technologies while upholding the integrity of natural and eco-friendly cosmetic offerings (Goyal & Jerold, 2023).

8.2.1.3 Natural actives that stimulates endogenous regeneration of NAD+

The use of natural actives that stimulate the body's endogenous regeneration of the crucial coenzyme NAD+ has emerged as a particularly promising approach in the development of innovative natural cosmeceuticals. Compounds such as nicotinamide riboside and nicotinic acid hold the ability to help maintain healthy cellular energy levels and promote skin rejuvenation by directly supporting the natural production of NAD+ within the body (Boo,2021). It represents a novel and promising avenue for the development of next-generation natural cosmeceuticals by harnessing body's own mechanisms for maintaining healthy NAD+ levels. These innovative natural actives can help support cellular energy metabolism, promote skin rejuvenation, and improve overall skin health. Incorporating these potent NAD+ precursors into advanced nanocarrier-based delivery systems can further enhance their bioavailability and overall efficacy, enabling the seamless delivery of the benefits of these natural ingredients while simultaneously improving product performance through the strategic application of nanotechnology. This synergistic combination of natural, biologically-active compounds and cutting-edge nanoscale delivery mechanisms holds immense potential to comprehensively address the growing consumer demand for sustainable, yet high-performing cosmetic solutions, all while effectively addressing any lingering regulatory concerns regarding the safety and efficacy of these sophisticated nanocosmetic formulations (Conlon, 2022).

While the targeting of NAD+ production through natural precursors like nicotinamide riboside and nicotinic acid may represent a novel and promising approach for next-generation natural cosmeceuticals, there are potential concerns that must be carefully considered. The use of these innovative natural actives, even when combined with advanced nanocarrier-based delivery systems, could raise regulatory challenges related to safety and efficacy. The seamless incorporation of these natural compounds into high-performance, sustainable cosmetic formulations may not fully address lingering regulatory concerns, as the complex interactions between the biological mechanisms, nanoparticle carriers, and cosmetic ingredients must be thoroughly evaluated. Additionally, the strategic synergy between natural, biologically-active ingredients and cutting-edge nanotechnology, while holding great promise, may face consumer skepticism regarding the perceived "naturalness" of these sophisticated nanocosmetic products. Careful balancing of consumer demand, industry innovation, and regulatory oversight will be crucial for ensuring the responsible development and adoption of these next-generation natural cosmeceuticals (Cornell et al., 2019).

8.2.1.4 Software and automation

The ability for manufacturers to utilize advanced software and prototyping tools now allows clients to visualize and modify product designs before production, ensuring that the final products align closely with specific client requirements. This increased customization has empowered clients to create unique products that cater directly to their preferences and needs, enhancing overall consumer satisfaction. Additionally, these advancements have streamlined the manufacturing process, leading to improved efficiency and greater flexibility in production. Furthermore, the integration of automation and smart manufacturing technologies has further transformed the cosmetic industry. Building upon the advancements in technology, automation has emerged as a pivotal force in the production and distribution of EOs and fragrances, fundamentally reshaping the cosmetics and perfumery industry. By implementing automation solutions, companies can significantly boost their operational efficiency, allowing them to streamline processes from extraction to packaging. This efficiency is crucial, especially as consumer demands evolve, necessitating quicker turnaround times and increased production capacity to cater to a diverse range of products (Neytal et al., 2024). Furthermore, automation plays a critical role in minimizing waste during production and distribution, which not only enhances sustainability efforts but also reduces operational costs. As companies navigate the complexities of compliance with various industry regulations, automation provides the necessary tools to ensure adherence to these standards, thereby mitigating risks associated with non-compliance. Automated systems can now handle repetitive tasks with precision, reducing the risk of human error and improving product consistency. The adoption of robotics and artificial intelligence has also enabled real-time monitoring and optimization of manufacturing processes, leading to enhanced quality control and faster turnaround times (Lee et al., 2015). Overall, the integration of automation in the EOs and fragrances sector not only optimizes production timelines but also contributes to a more responsible and responsive manufacturing landscape.

These innovations have also enabled the cosmetic industry to better cater to the evolving preferences of conscious consumers. Automated systems and data analytics can help manufacturers identify emerging trends and tailor their product offerings accordingly, ensuring they remain competitive and responsive to the market. Moreover, the integration of sustainable manufacturing practices, such as reduced waste and energy consumption, has become a key priority for many cosmetic companies, driven by consumer demand for eco-friendly products. By leveraging technology and data-driven insights, the industry can develop innovative cosmetic formulations that not only meet consumer needs but also align with broader sustainability goals, further enhancing brand reputation and customer loyalty (Feng et al., 2018).

While the cosmetic industry has indeed leveraged innovations to cater to conscious consumers, there are concerns that such approaches may not always deliver on their intended sustainability and environmental benefits. Automated systems and data analytics, while enabling responsiveness to market trends, could also contribute to increased production and resource consumption, potentially negating the positive impact of eco-friendly practices. Moreover, the push for innovative formulations that align with sustainability goals may overlook the limitations of current technologies or the trade-offs involved in balancing consumer demands with environmental considerations. As the industry continues to evolve, it will be crucial to carefully evaluate the long-term implications of these innovations and ensure that they truly deliver on their promises of sustainability and environmental stewardship, rather than merely serving as marketing tactics (Huang, 2021).

8.2.1.5 Digital tools

In the rapidly evolving landscape of cosmetics, digital tools have played a pivotal role in reshaping marketing strategies, particularly for products containing EOs. One of the most significant transformations has been the rise of user-generated content through social media, which has significantly boosted awareness and visibility for cosmetic brands that utilize EOs. This shift allows consumers to share their experiences and recommendations, creating a more authentic connection with potential buyers. Furthermore, community building on digital platforms enables brands to engage directly with customers who are particularly interested in EO cosmetics, fostering loyalty and trust among their audience. The influence of social media content creators cannot be understated, as they have made eco-friendly cosmetic products more accessible and appealing to a broader audience, often showcasing the benefits of EOs in their routines. In conjunction with this, targeted social media advertising has enhanced the promotion of these eco-friendly products, allowing brands to reach specific demographics effectively and efficiently (Pawar & Pawar, 2023). Additionally, digital marketing techniques help brands maximize their profits by strategically promoting EO-based offerings, capitalizing on the growing consumer interest in natural and sustainable products. To further bolster their online presence, brands are employing search engine optimization strategies, ensuring that their EO cosmetics appear prominently in search results, thereby increasing visibility and attracting new customers. Overall, the integration of digital tools into marketing strategies has allowed cosmetic brands to not only adapt to consumer preferences but also thrive in a competitive market (Constantinides et al., 2009).

While digital tools have played a significant role in reshaping marketing strategies for EO-based cosmetic products, there are valid concerns that must be addressed. The rise of user-generated content on social media, though boosting awareness and visibility, can also lead to the spread of misinformation and unsubstantiated claims about the benefits of these products. The influence of social media content creators, while making eco-friendly cosmetics more accessible, may not always provide accurate or reliable information, potentially misleading consumers. Furthermore, the targeted social media advertising and digital marketing techniques employed by brands could be perceived as manipulative, prioritizing profit maximization over consumer education and transparency. While digital tools have enabled cosmetic brands to adapt to consumer preferences, there is a risk that these strategies could compromise the integrity of the natural and sustainable claims associated with EO-based offerings. Careful consideration must be given to striking a balance between leveraging digital platforms and ensuring that consumers are making informed decisions based on scientific evidence and responsible marketing practices (Gcora et al., 2019).

8.3 REGULATORY RESPONSES TO INNOVATIONS

The cosmetic industry has seen notable regulatory shifts in response to innovations in EOs and fragrances, reflecting a broader trend toward enhanced consumer safety and environmental protection. The innovation in the use of EOs and fragrances in cosmetic products has been a topic of significant interest and concern within the regulatory landscape. The growing awareness of the potential risks posed by synthetic preservatives and fragrances has led to an increased demand for more natural alternatives with better efficacy, which leads to innovation in the field (Rathore et al., 2023). This expansion has been driven by consumer demand for innovative and natural products, particularly those incorporating EOs, which are perceived as safer and more effective alternatives to synthetic fragrances. However, the regulatory bodies tasked with ensuring the safety and efficacy of cosmetic products have faced challenges in keeping pace with the evolving market and the complexities of natural ingredients and new technologies.

Regulatory agencies, such as the US Food and Drug Administration (FDA) and the European Union's Cosmetic Regulation, have implemented various measures to address the development and innovation of new EO and fragrance technologies and their uses in cosmetic products as shown in Figure 8.2. These measures include establishing guidelines for the safe use of these ingredients, requiring specific labeling and disclosure of ingredients, and conducting safety assessments to evaluate the potential risks associated with their use (Lauriola & Corazza, 2019). Another pivotal example of this is the European Union's REACH regulation, which requires extensive safety testing and registration of chemicals used in fragrances, thereby imposing stringent standards on manufacturers. This regulation not only ensures that harmful substances are identified and mitigated but also emphasizes transparency in ingredient disclosure, allowing consumers to make informed choices about the products they use.

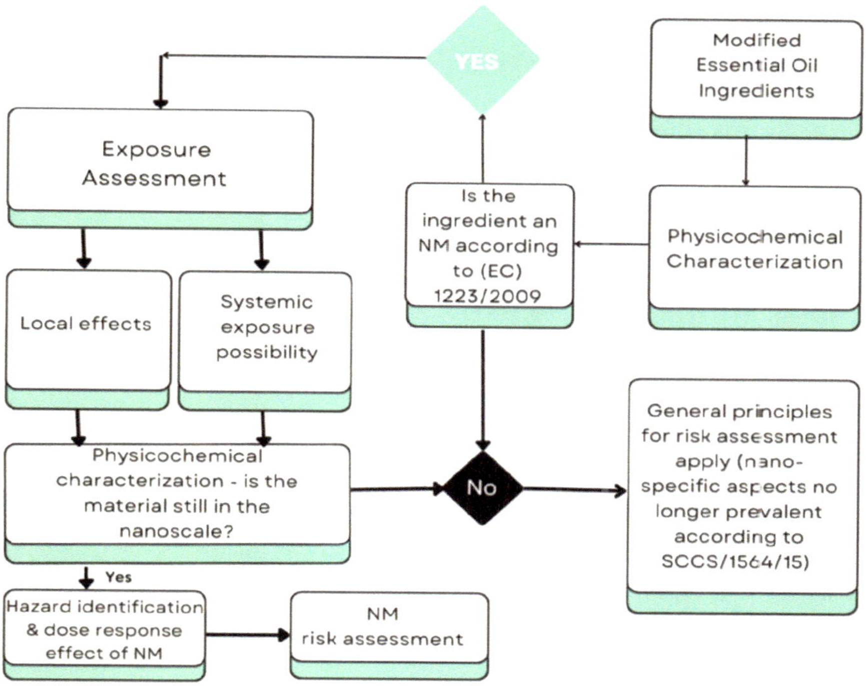

Figure 8.2 Regulation of cosmetic products containing essential oil and fragrances in nano size or using nanomaterial.

The regulatory landscape continues to evolve as new research and scientific information on the effects of EOs and fragrances emerge. The regulatory bodies worldwide are increasingly tightening rules surrounding fragrance ingredients, driven by growing pressure from consumers and advocacy groups who demand safer and more sustainable cosmetic (Bampidis et al., 2024; Es et al., 2017). This consumer advocacy is reshaping the industry, prompting companies to reformulate products to comply with new regulations while also prioritizing the use of natural and eco-friendly ingredients. As a result, the interplay between regulatory frameworks and consumer expectations is fostering a more responsible approach to the use of EOs and fragrances in cosmetics, ultimately benefiting both public health and environmental sustainability.

As the demand for natural and sustainable cosmetics grows, this regulatory evolution becomes even more imperative. This ongoing process is crucial, as it allows regulatory bodies to stay up-to-date and make informed, evidence-based decisions. These decisions enable the cosmetics industry to formulate products that are not only safe and effective but also responsive to consumer preferences for more natural and sustainable options. At the same time, the regulatory measures play a vital role in striking a delicate balance—protecting public health and safety while also fostering industry innovation and empowering consumers to make well-informed choices about the products they use. This balanced approach is crucial for upholding public trust, supporting industry progress, and promoting the availability of high-quality, environmentally-conscious cosmetic products (Bialas et al., 2023).

However, the evolving regulatory landscape has faced criticism from some who argue that the measures implemented by agencies like the FDA and EU Cosmetic Regulation are overly restrictive and stifle industry innovation. They contend that the extensive safety assessments and disclosure requirements place an undue burden on cosmetics manufacturers, especially smaller companies, and limit their ability to bring new natural and sustainable products to market in a timely manner (Wells,2023). Critics also suggest that the regulatory focus on potential risks associated with EOs and fragrances ignores the potential benefits these natural ingredients can offer, and that a more balanced approach is needed to foster both consumer safety and industry progress. As the demand for natural cosmetics continues to grow, this tension between regulatory oversight and industry flexibility remains a subject of ongoing debate. The ongoing debate surrounding the regulatory responses to the development and innovation of new EO and fragrance technologies and their uses in cosmetic products highlights the complex and multifaceted nature of this issue.

Proponents of a more balanced approach suggest that a deeper consideration of the potential benefits of EOs and fragrances, in addition to the potential risks, could lead to a more nuanced and flexible regulatory framework (Barbieri & Borsotto, 2018). This could involve a more collaborative process between regulatory bodies, industry stakeholders, and scientific experts to develop guidelines that accommodate innovation while prioritizing consumer health and safety.

8.3.1 Variation in regulation across different regions

The regulation of EOs and fragrances in cosmetic products varies significantly across different regions, reflecting diverse approaches to consumer safety and product efficacy. In the United States, there is no overarching federal regulation specifically dedicated to cosmetics, which creates a complex landscape for manufacturers and consumers alike. Instead, the primary laws governing cosmetics include the Federal Food, Drug, and Cosmetic Act (FD&C Act) and the Fair Packaging and Labeling Act (FPLA), which provide a framework for safety and labeling requirements. The Cosmetic Ingredient Review (CIR), an expert panel established by the Personal Care Products Council (PCPC), plays a crucial role in evaluating the safety of cosmetic ingredients, including EOs (Barbieri & Borsotto, 2018). While the FDA oversees cosmetic products, it does not require pre-market approval, allowing companies significant leeway in the use of EOs, provided they come from plants permitted for such use. However, certain naturally occurring substances in EOs, such as safrole, methyleugenol, and furocoumarins, face strict regulatory limits to mitigate potential health risks associated with their use in finished cosmetic products. Furthermore, the Cosmetics Regulation in the European Union takes a more stringent approach, as outlined in Annex II, which explicitly prohibits 92 plant species and their derivatives from being used in cosmetics, underscoring a precautionary principle in cosmetic safety regulation (Morel et al., 2023). This divergence highlights the need for consumers and manufacturers to be well-informed about the specific regulations that govern the use of EOs and fragrances in their respective regions.

The variations in cosmetic regulations across different countries significantly impact international trade, creating both challenges and opportunities for businesses. For example, while the United States relies on the CIR for ingredient evaluation, the European Union has a more stringent regulatory framework that requires comprehensive safety assessments and pre-market approvals, affecting the market flexibility for companies operating in both regions. The classification of products can also differ, leading to distinct regulatory requirements that complicate compliance for manufacturers aiming to market their products globally (Ferreira et al., 2022). Moreover, labeling requirements can vary greatly; the United States has specific rules regarding claims about ingredients like SPF, while other countries emphasize transparency and consumer protection in labeling practices. These discrepancies can hinder trade by creating barriers to entry for companies that must navigate multiple regulatory landscapes and ensure compliance with diverse labeling and safety standards (Manful et al., 2024). As the global market becomes increasingly interconnected, there is a recognized need for harmonized regulations that facilitate international trade, particularly in the cosmetics sector. The recent shift in China's policies, removing the mandatory animal testing requirement for general cosmetics while still maintaining specific preconditions, exemplifies ongoing changes in global regulatory attitudes that could reshape trade dynamics further. Thus, understanding these regulatory differences is crucial for companies looking to operate successfully in international markets, as they must adapt their strategies to meet varying compliance demands across regions.

8.3.2 Role of governmental and non-governmental organizations

In the realm of fragrance regulation, both governmental and non-governmental organizations play pivotal roles in establishing and enforcing safety standards. For instance, the European Union has implemented comprehensive regulatory frameworks like REACH, which are designed to mitigate the risks associated with chemical substances, including fragrances, underscoring the active involvement of government in regulation enforcement. Similarly, various countries have established specific regulations governing the use of fragrances in consumer products, illustrating the essential role governmental bodies play in maintaining public safety (Chávez-Hernández et al., 2024). Complementing these governmental efforts, non-governmental organizations such as the IFRA contribute significantly through the development of standards that address the safe usage of fragrance ingredients and labeling requirements. This collaborative approach between industry self-regulation and governmental oversight ensures that safety standards are upheld, providing consumers with confidence in the products they use (Lee et al., 2024). Furthermore, organizations like the US FDA and the FTC are tasked with enforcing regulations such as the FPLA, which aims to protect consumers by requiring clear labeling and preventing misleading packaging practices. It emphasizes key provisions that mandate the disclosure of product identity and manufacturer information, reinforcing transparency in consumer products. However, the regulatory landscape is continuously evolving, presenting challenges for regulatory departments as they adapt to new requirements. This highlights the necessity for ongoing collaboration between governmental bodies and non-governmental organizations, fostering dialogue that advocates for practical regulations and realistic timelines for compliance. Non-governmental organizations also play a vital role in supporting their members through active regulatory monitoring, ensuring that companies remain informed about compliance obligations and best practices in the industry. Collectively, these organizations contribute to a comprehensive regulatory framework that safeguards consumer and environmental health while promoting fair market practices (Azam, 2024).

8.3.3 Controversies and public perception

Despite the rising popularity of EOs in the cosmetic industry, several controversies surrounding their safety and efficacy persist. For instance, certain fragrance materials, which can include EOs, are linked to allergies and sensitivities, raising concerns about their use in consumer products. The IFRA has taken a proactive stance by prohibiting the inclusion of flagged materials known for potential risks, including those associated with reproductive and developmental toxicity (Dallay et al., 2024). Moreover, while EOs are often perceived as safe and natural alternatives, this belief may be misleading due to underlying safety issues that can arise from their use. Recent studies have highlighted the complexities of EOs' safety profiles, indicating that the debate surrounding their use is ongoing and multifaceted. These studies have revealed that EOs can potentially cause skin irritation, sensitization, and even reproductive and developmental toxicity, in addition to potential antimicrobial, antifungal, and antioxidant properties. Compounding these concerns is the issue of standardization; EOs are notoriously difficult to standardize due to variations in growing conditions, genetic differences among plants, and methods of harvesting, which can all impact their potency and safety. As a result, the cosmetic industry faces significant challenges in ensuring that these products are not only effective but also safe for consumers, prompting a careful reevaluation of how EOs and their components are integrated into formulations (Aswandi et al., 2024).

One of the key factors contributing to the controversies surrounding the use of EOs in cosmetics is the complex and often contradictory research on their potential benefits and risks. While some studies have reported antimicrobial, antifungal, and antioxidant properties of EOs, others have highlighted their potential for causing skin irritation, sensitization, and even reproductive and developmental toxicity (Sarkic & Stappen, 2018; Sharmeen et al., 2021) This has led to a heightened awareness of the need for more rigorous safety assessments and regulatory frameworks to ensure the responsible and safe use of these natural ingredients.

The issue of standardization is also a significant challenge in the cosmetic industry's use of EOs. Variations in growing conditions, genetic differences among plants, and methods of harvesting can all impact the potency and safety of EOs, making it difficult to ensure consistent quality and safety across products (Salgueiro et al., 2010). Efforts to address this challenge have included the development of more robust analytical methods for identifying and quantifying the active compounds in EOs, as well as the establishment of industry guidelines and regulations to ensure the quality and safety of these ingredients.

To address these concerns, regulatory bodies and industry organizations have taken steps to develop more comprehensive frameworks for the use of EOs and fragrances in cosmetic products. For example, the European Union has implemented the Cosmetic Products Regulation, which requires detailed safety assessments and disclosure of fragrance ingredients to consumers (Commission, 1999; Klaschka, 2016). Similarly, the IFRA has established a set of voluntary standards and guidelines for the safe use of fragrance materials, including EOs, in consumer products. These initiatives aim to strike a balance between innovation, consumer safety, and transparency, as the cosmetic industry continues to navigate the complexities surrounding the use of these natural ingredients.

8.3.3.1 Public perception

In recent years, public perception regarding product ingredients has undergone a significant transformation, reflecting a growing demand for transparency and safety. Since 1999, consumer attitudes towardS product information have shifted remarkably, with individuals now seeking comprehensive details about the contents of the products they use. This heightened awareness is particularly evident in the realm of personal care and cosmetics, where consumers are increasingly skeptical about synthesized chemicals and their potential negative effects. For instance, when Johnson & Johnson acquired Aveeno, there was little interest among consumers in understanding the specifics of product contents, but today, the narrative has changed dramatically, as consumers are more inquisitive and concerned about what they apply to their bodies. This shift has prompted companies to respond to consumer demands for better, simpler, and more natural ingredients, leading to noteworthy changes in product formulations. Major retailers, such as Costco and Whole Foods, are also attuned to these preferences, influencing food producers to adapt their offerings in response to consumer concern over ingredients. Additionally, the removal of harmful chemicals like formaldehyde from baby products by Johnson & Johnson showcases how public scrutiny can drive corporate decision-making, further illustrating the impact of consumer sentiment on market trends (Burger et al., 2019). As consumers express their opinions vocally through social media, the natural personal care industry is projected to experience continued growth, reflecting a favorable public perception that prioritizes health and safety over conventional product offerings [6]. This evolving landscape underscores a fundamental shift in consumer behavior toward a more informed and proactive approach to product selection.

In summary, the use of EOs and fragrances in cosmetic products is a complex and multifaceted issue. While these natural ingredients offer potential benefits, there are also significant safety concerns that need to be addressed through rigorous safety assessments and comprehensive regulatory frameworks. The issue of standardization further complicates the responsible integration of EOs into cosmetic formulations, requiring the development of robust analytical methods and industry guidelines. As public perception continues to evolve, with consumers demanding more transparency and safer products, the cosmetic industry faces the challenge of balancing innovation, safety, and consumer expectations. Addressing these concerns will be crucial for ensuring the responsible and sustainable use of EOs and fragrances in cosmetic products moving forward (Ferreira et al., 2022; Sharmeen et al., 2021).

The shifting public perception toward product ingredients has also added to the challenges faced by the cosmetic industry (Duran et al., 2014). Consumers are now more skeptical about synthesized chemicals and are demanding greater transparency and safety in the products they use. This has prompted companies to respond by reformulating their products with better, simpler, and more natural ingredients, reflecting the growing preference for health and safety over conventional product offerings. As consumer sentiment continues to evolve, the cosmetic industry must adapt and address the concerns of the public to ensure the responsible and sustainable use of EOs and fragrances in their products (Singh et al., 2019).

8.4 CONSUMER SAFETY CONCERNS

The increasing popularity of EOs in cosmetics and personal care products raises important consumer safety concerns that necessitate careful consideration. While the majority of the population can safely use EOs, as indicated by studies showing low incidence of allergic reactions, there are still significant risks for susceptible individuals, particularly those with pre-existing skin conditions or sensitivities. Research indicates that 79 different EOs have been associated with allergic reactions, underscoring the need for vigilance in product formulation. the declaration of specific allergens that can arise from EOs, as there are 26 identified allergenic fragrances, with 18 present in EOs, which highlights the importance of proper labeling to mitigate risks for sensitive consumers (Sarkic & Stappen, 2018).

Many consumers are drawn to the perceived benefits of natural cosmetics, often believing that they are inherently safer and less harmful than their synthetic counterparts. However, this perception may not always be accurate, as the allergenic potential of EO and fragrance ingredients in these products can be just as concerning (Klaschka, 2016). They can develop conditions like allergenicity, irritation risks, sensitization, and contact dermatitis. Consumers may be misled by the attractive packaging, appealing marketing claims, and lack of warning symbols on these products, leading them to trust that they are safe, even if they contain potentially hazardous substances.

The use of natural ingredients in cosmetics, including the use of sustainable agriculture and organic farming, has been a significant challenge for the industry in recent years. This has been compounded by regulatory changes, product safety concerns, and increasing consumer demand for scientific data to support product claims. The cosmetic industry has had to navigate this complex landscape, balancing the desire for natural and sustainable ingredients with the need to ensure the safety and efficacy of their products (Goyal & Jerold, 2023; Panico et al., 2019).

At the same time, proponents of natural and organic cosmetic products argue that these products offer significant benefits that outweigh any potential safety concerns. They contend that the use of synthetic chemicals and preservatives in traditional cosmetics poses greater long-term health risks, and that the natural ingredients in these products are inherently safer and more sustainable (Nduka et al., 2019). Furthermore, they assert that the increased transparency and disclosure of ingredients in natural cosmetics allows consumers to make more informed choices about the products they use. While the allergenic potential of fragrance ingredients is a valid concern, these advocates maintain that the overall safety profile of natural cosmetics is superior to that of their synthetic counterparts.

The safety and efficacy of natural cosmetic products is likely to continue as the industry navigates the complex landscape of regulatory changes, product safety concerns, and increasing consumer demand for scientific data to support product claims. This balance between the desire for natural and sustainable ingredients, and the need to ensure the safety and efficacy of cosmetic products remains a significant challenge for the industry (Fonseca-Santos et al., 2015). Proponents of natural and organic cosmetic products argue that these products offer significant benefits that outweigh any potential safety concerns, while critics contend that the allergenic potential of fragrance ingredients in these products can be just as concerning as the use of synthetic chemicals in traditional cosmetics. As consumer demand for natural and sustainable products continues to grow, the industry will need to find ways to address these concerns and provide consumers with safe and effective cosmetic options.

8.4.1 Debates over natural versus synthetic ingredients

The debate over natural versus synthetic ingredients in cosmetic products is a complex and multifaceted issue that requires careful consideration of various factors. Proponents of natural cosmetics argue that natural ingredients are inherently safer and more sustainable, as they are derived from plant-based sources and do not involve the use of synthetic chemicals. They contend that the natural compounds found in EOs and fragrances are less likely to cause adverse reactions or long-term health concerns compared to the synthetic counterparts used in traditional cosmetics (Irshad et al., 2020).

However, critics of natural cosmetics point out that the allergenic potential of fragrance compounds, including those derived from natural sources, can be just as concerning as the use of synthetic chemicals (Bucheli et al., 2018). They argue that the increased reliance on natural ingredients in cosmetic formulations does not necessarily guarantee greater safety, as some natural compounds can also trigger allergic responses in sensitive individuals. Furthermore, the lack of standardized regulations and scientific data to support the safety and efficacy claims of natural cosmetics can be a significant challenge for consumers trying to make informed choices.

Ultimately, the debate over natural versus synthetic ingredients in cosmetics highlights the need for a balanced and nuanced approach that prioritizes consumer safety and transparency. Cosmetic manufacturers must strive to provide consumers with clear and accurate information about the ingredients used in their products, regardless of whether they are natural or synthetic (Fung et al., 2018). Additionally, ongoing research and regulatory oversight are crucial to ensuring that both natural and synthetic cosmetic products meet rigorous safety standards and deliver the promised benefits to consumers. As the cosmetic industry continues to evolve, the resolution of this debate will require a collaborative effort among industry, regulators, and consumers to find the best solutions that prioritize safety, sustainability, and consumer empowerment.

8.4.2 Role of sustainability in the development of new essential oil and fragrance technologies

The sustainability plays a critical role in shaping the future of EO and fragrance technologies, as it emphasizes the use of natural and renewable resources in production. As the fragrance industry pivots toward eco-friendly practices, the development of sustainable fragrance oils becomes a focal point, aligning with consumer demand for environmentally responsible products as shown in Figure 8.3. This shift not only reflects a commitment to sustainability but also influences the methodologies of harvesting and extraction processes, ensuring that these practices are both ethical and efficient. Moreover, certification is becoming increasingly significant, as consumers seek assurance that the products they purchase meet high sustainability standards. While

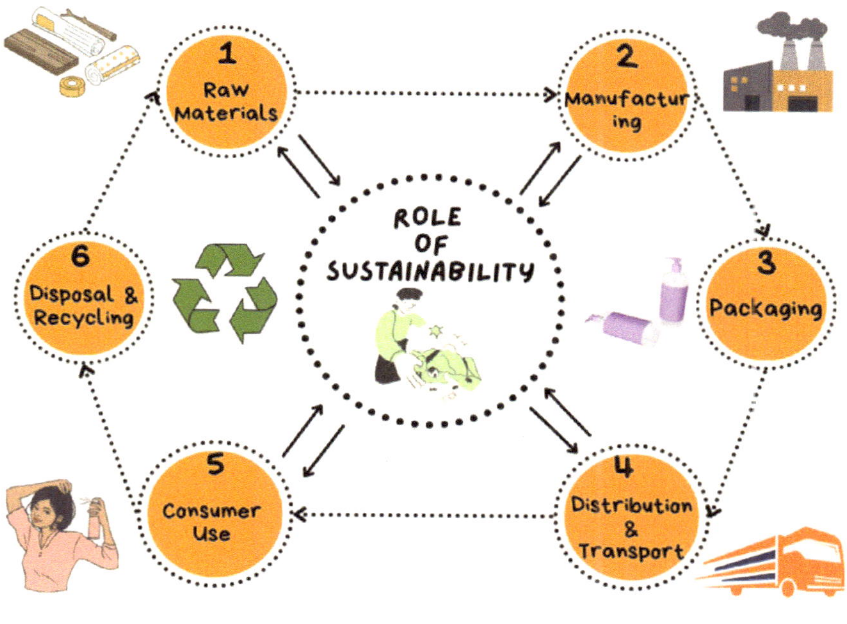

Figure 8.3 Role of sustainability in cosmetic industry.

the fragrance sector lags behind other cosmetics categories, such as hair and skincare, in terms of certification prevalence, it is still vital for establishing credibility in sustainable sourcing. For example, membership in organizations like the Natural Resources Stewardship Circle (NRSC) enhances the reputation of producers committed to sustainable practices. Additionally, obtaining certifications such as Organic, Fair Wild, or Fair Trade allows companies to demonstrate their commitment to ethical production, although certified sustainable perfumes remain relatively rare, highlighting a notable gap in the market. This gap indicates a burgeoning opportunity for brands to cater to the growing consumer interest in sustainably produced EOs, which are increasingly being sought after for their ethical qualities in cosmetic applications (Sharmeen et al., 2021). As consumers become more invested in the stories behind the products they use, the demand for transparently sourced and fairly traded EOs is likely to increase, shaping the future landscape of fragrance technology and production.

REFERENCES

11 Gorgeous Perfumes Made from Essential Oils. (n.d.). Retrieved August 27, 2024, from https://www.instyle.com/perfumes-from-essential-oils-8647488

Aburjai, T., & Natsheh, F. M. (2003). Plants used in cosmetics. *Phytotherapy Research: An International Journal Devoted to Pharmacological and Toxicological Evaluation of Natural Product Derivatives*, *17*(9), 987–1000.

Amazon.in : hyaluronic acid moisturizer. (n.d.). Retrieved August 27, 2024, from https://www.amazon.in/hyaluronic-acid-moisturizer/s?k=hyaluronic+acid+moi sturizer

Aswandi, A., Kholibrina, C. R., & Kuspradini, H. (2024). Essential oils for cosmetics application. In *Biomass-based cosmetics: Research trends and future outlook* (pp. 151–173). Springer.

Azam, U. (2024). A new era in cosmetology. *European Journal of Biomedical*, *11*(5), 91–102.

Baki, G., & Peeters, M. J. (2019). Exploring the impact of technology use with cosmetic science guest-speakers: A qualitative study. *INNOVATIONS in Pharmacy*, *10*(2).

Bampidis, V., Azimonti, G., Bastos, M. L., Christensen, H., Durjava, M., Kouba, M., López-Alonso, M., López Puente, S., & Marcon, F. on Additives, E. P., or Substances used in Animal Feed (FEEDAP), P, & others. (2024). Safety and efficacy of a feed additive consisting of an essential oil derived from the leaves of *Cymbopogon nardus* (L.) Rendle (citronella oil) for use in all animal species (FEFANA asbl). *EFSA Journal*, *22*(5), e8790.

Barbieri, C., Borsotto, P., & others. (2018). Essential oils: Market and legislation. *Potential of Essential Oils*, 107–127.

Berger, R. G. (2007). *Flavours and fragrances: Chemistry, bioprocessing and sustainability.* Springer Science+Business Media.

Bialas, I., Zelent-Kraciuk, S., & Jurowski, K. (2023). The skin sensitisation of cosmetic ingredients: Review of actual regulatory status. *Toxics*, *11*(4), 392.

Boo, Y. C. (2021). Mechanistic basis and clinical evidence for the applications of nicotinamide (niacinamide) to control skin aging and pigmentation. *Antioxidants*, *10*(8), 1315.

Bucheli, T. D., Strobel, B. W., & Hansen, H. C. B. (2018). Personal care products are only one of many exposure routes of natural toxic substances to humans and the environment. *Cosmetics*, *5*(1), 10.

Burger, P., Plainfossé, H., Brochet, X., Chemat, F., & Fernandez, X. (2019). Extraction of natural fragrance ingredients: History overview and future trends. *Chemistry\& Biodiversity*, *16*(10), e1900424.

Butschke, A., Droß, A., Dünnebier, K., Laube, I., & Weiler, A. (2016). Experiences and statistical evaluation of serious undesirable effects of cosmetic products in the EU. *Cosmetics*, *3*(3), 25.

Carvalho, I. T., Estevinho, B. N., & Santos, L. (2016). Application of microencapsulated essential oils in cosmetic and personal healthcare products–a review. *International Journal of Cosmetic Science*, *38*(2), 109–119.

Chadha, S. (2021). Recent advances in nano-encapsulation technologies for controlled release of biostimulants and antimicrobial agents. In *Advances in nano-fertilizers and nano-pesticides in agriculture: A smart delivery system for crop improvement* (pp. 29–55). https://doi.org/10.1016/B978-0-12-820092-6.00002-1

Chauhan, A., & Chauhan, C. (2023). Emerging trends of nanotechnology in beauty solutions: A review. *Materials Today: Proceedings*, *81*, 1052–1059.

Chávez-Hernández, J. A., Velarde-Salcedo, A. J., Navarro-Tovar, G., & Gonzalez, C. (2024). Safe nanomaterials: From their use, application and disposal to regulations. *Nanoscale Advances*.

Commission, E. (1999). The rules governing cosmetic products in the European Union. *Cosmetics Legislation*, *1*.

Conlon, N. J. (2022). The role of NAD+ in regenerative medicine. *Plastic and Reconstructive Surgery*, *150*, 41S–48S.

Constantinides, E., Romero, C. L., & Boria, M. A. G. (2009). Social media: A new frontier for retailers? In *European retail research* (pp. 1–28).

Cornell, E. M., Janetos, T. M., & Xu, S. (2019). Time for a makeover-cosmetics regulation in the United States. *Journal of Cosmetic Dermatology*, *18*(6), 2041–2047.

Dallay, C., Malhiac, C., Picard, C., & Savary, G. (2024). Fragrance in dermocosmetic emulsions: From microstructure to skin application. *International Journal of Cosmetic Science*, *46*(1), 1–23.

Dubey, S. K., Dey, A., Singhvi, G., Pandey, M. M., Singh, V., & Kesharwani, P. (2022). Emerging trends of nanotechnology in advanced cosmetics. *Colloids and Surfaces B: Biointerfaces*, *214*, 112440.

Duran, I., Bikfalvi, A., & Llach, J. (2014). New facets of quality. A multiple case study of green cosmetic manufacturers. A multiple case study of green cosmetic manufacturers (September 5, 2014). *European Accounting and Management Review*, *1*(1).

Es, I., Khaneghah, A. M., & Akbariirad, H. (2017). Global regulation of essential oils. In *Essential oils in food processing: Chemistry, safety and applications* (pp. 327–338).

EWG. (n.d.). *Products That Contain Polycitronellol __ Skin Deep® Cosmetics Database _ EWG*. Retrieved August 23, 2024, from https://www.ewg.org/skindeep/browse/ingredients/701307-CHITIN_EXTRACT/

Feng, C., Chen, H., & Ho, J. C. (2018). Promoting the diffusion of sustainable innovations in the cosmetic industry. *2018 IEEE Technology and Engineering Management Conference (TEMSCON)*, 1–5.

Ferreira, M., Matos, A., Couras, A., Marto, J., & Ribeiro, H. (2022). Overview of cosmetic regulatory frameworks around the world. *Cosmetics*, *9*(4), 72.

Fonseca-Santos, B., Corrêa, M. A., & Chorilli, M. (2015). Sustainability, natural and organic cosmetics: Consumer, products, efficacy, toxicological and regulatory considerations. *Brazilian Journal of Pharmaceutical Sciences*, *51*(1), 17–26.

Fung, E. S., Drechsel, D. A., Towle, K. M., Hoang, M. T., Novick, R. M., Poteete, C., Paustenbach, D. J., & Monnot, A. D. (2018). Screening-level safety assessment of personal care product constituents using publicly available data. *Cosmetics*, *5*(2), 38.

Fytianos, G., Rahdar, A., & Kyzas, G. Z. (2020). Nanomaterials in cosmetics: Recent updates. *Nanomaterials*, *10*(5), 979.

Galea, C., Cocoş, D. I., Feier, F., & Moales, D. (2023). The use of essential oils in the development of dermato-cosmetic products. *Medicine and Materials*, *3*(1), 31–36.

Gcora, N., Maoneke, P. B., & Isabirye, N. (2019). A model to enhance the perceived trustworthiness of small and medium enterprises selling natural essential oils through e-marketplaces. In *New insights on trust in business-to-business relationships: A multi-perspective approach* (pp. 37–52). Emerald Publishing Limited.

Gomes, C., Silva, A. C., Marques, A. C., Sousa Lobo, J., & Amaral, M. H. (2020). Biotechnology applied to cosmetics and aesthetic medicines. *Cosmetics*, *7*(2), 33.

Goyal, N., & Jerold, F. (2023). Biocosmetics: Technological advances and future outlook. *Environmental Science and Pollution Research*, *30*(10), 25148–25169.

Gupta, P. L., Rajput, M., Oza, T., Trivedi, U., & Sanghvi, G. (2019). Eminence of microbial products in cosmetic industry. *Natural Products and Bioprospecting*, *9*, 267–278.

Gupta, V., Mohapatra, S., Mishra, H., Farooq, U., Kumar, K., Ansari, M. J., Aldawsari, M. F., Alalaiwe, A. S., Mirza, M. A., & Iqbal, Z. (2022). Nanotechnology in cosmetics and cosmeceuticals—a review of latest advancements. *Gels*, *8*(3), 173.

Guzmán, E., & Lucia, A. (2021). Essential oils and their individual components in cosmetic products. *Cosmetics*, *8*(4), 114.

Hooda, A., Sradhanjali, M., & others. (2017). Formulation and evaluation of novel solid lipid microparticles for the sustained release of ofloxacin. *Pharmaceutical Nanotechnology*, *5*(4), 329–341.

Hosoi, J., Koyama, J., & Ozawa, T. (2017). New aspects of cosmetics and cosmetic science. In Sakamoto, K., Lochhead, R. Y., Maibach, H. I., Yamashita, Y. (Eds.), *Cosmetic science and technology: Theoretical principles and applications* (pp. 87–100).

Huang, Y. (2021). Technology innovation and sustainability: Challenges and research needs. In *Clean technologies and environmental policy* (Vol. 23, pp. 1663–1664). Springer.

Hyaluronic acid: Uses, side effects, and risks. (n.d.). Retrieved August 29, 2024, from https://www.medicalnewstoday.com/articles/326385

Irshad, M., Subhani, M. A., Ali, S., & Hussain, A. (2020). Biological importance of essential oils. *Essential Oils-Oils of Nature*, *1*, 37–40.

Kan, H., Chou, X., Li, M., Xi, C., Zhou, T., Xiao, Z., & Ding, Y. (2024). Enhancing the chemoselective hydrogenation of o-chloronitrobenzene over nanoporous $Ni/NiAl_3$ catalysts. *Applied Surface Science*, *649*, 159120.

Kaul, S., Gulati, N., Verma, D., Mukherjee, S., & Nagaich, U. (2018). Role of nanotechnology in cosmeceuticals: A review of recent advances. *Journal of Pharmaceutics*, *2018*(1), 3420204.

Klaschka, U. (2016). Natural personal care products—Analysis of ingredient lists and legal situation. *Environmental Sciences Europe*, *28*, 1–14.

Lauriola, M. M., & Corazza, M. (2019). The wild market of natural cosmetics of obscure safety. *Dermatology*, *235*(6), 527–528.

Lee, S. L., O'Connor, T. F., Yang, X., Cruz, C. N., Chatterjee, S., Madurawe, R. D., Moore, C. M. V., Yu, L. X., & Woodcock, J. (2015). Modernizing pharmaceutical manufacturing: From batch to continuous production. *Journal of Pharmaceutical Innovation*, *10*, 191–199.

Lee, I., Scrochi, C., Chon, O., Cancellieri, M. A., Ghosh, A., O'Brien, J., Ring, B., McNamara, C., & Api, A. M. (2024). Detailed aggregate exposure analysis shows that exposure to fragrance ingredients in consumer products is low: Many orders of magnitude below thresholds of concern. *Regulatory Toxicology and Pharmacology*, *148*, 105569.

López-Hortas, L., Flórez-Fernández, N., Torres, M. D., Ferreira-Anta, T., Casas, M. P., Balboa, E. M., Falqué, E., & Dom\'\inguez, H. (2021). Applying seaweed compounds in cosmetics, cosmeceuticals and nutricosmetics. *Marine Drugs*, *19*(10), 552.

Manful, M. E., Ahmed, L., & Barry-Ryan, C. (2024). Cosmetic formulations from natural sources: Safety considerations and legislative frameworks in the European Union. *Cosmetics*, *11*(3), 72.

Morel, S., Sapino, S., Peira, E., Chirio, D., & Gallarate, M. (2023). Regulatory requirements for exporting cosmetic products to extra-EU countries. *Cosmetics*, *10*(2), 62.

Nanoparticles found in 10 top brand cosmetics – Emerging Technology. (n.d.). Retrieved August 23, 2024, from https://emergingtech.foe.org.au/nanoparticles-found-10-top-brand-cosmetics/

Nduka, J. K., Kelle, H. I., & Odiba, I. O. (2019). Review of health hazards and toxicological effects of constituents of cosmetics. In *Poisoning in the modern world-new tricks for an old dog* (pp. 59–76).

Neytal, P., Maji, N., & Maji, S. (2024). Advances in cosmetic products towards a new future. In *preserving health, preserving earth: The path to sustainable healthcare* (pp. 193–214). Springer.

Panico, A., Serio, F., Bagordo, F., Grassi, T., Idolo, A., De Giorgi, M., Guido, M., Congedo, M., & De Donno, A. (2019). Skin safety and health prevention: An overview of chemicals in cosmetic products. *Journal of Preventive Medicine and Hygiene*, *60*(1), E50.

Pawar, A., & Pawar, A. (2023). Crafting effective marketing strategies for global success in the beauty and cosmetic industry. *IJO-International Journal of Business Management (ISSN 2811-2504)*, *6*(12), 14–24.

Rathore, S., Mukhia, S., Kumar, R., & Kumar, R. (2023). Essential oil composition and antimicrobial potential of aromatic plants grown in the mid-hill conditions of the Western Himalayas. *Scientific Reports*, *13*(1), 4878.

Salgueiro, L., Martins, A. P., & Correia, H. (2010). Raw materials: The importance of quality and safety. A review. *Flavour and Fragrance Journal*, *25*(5), 253–271.

Santos, A. C., Marto, J., Chá-Chá, R., Martins, A. M., Pereira-Silva, M., Ribeiro, H. M., & Veiga, F. (2022). Nanotechnology-based sunscreens—a review. *Materials Today Chemistry*, *23*, 100709. https://doi.org/10.1016/J.MTCHEM.2021.100709

Sarkic, A., & Stappen, I. (2018). Essential oils and their single compounds in cosmetics—A critical review. *Cosmetics*, *5*(1), 11.

Sharmeen, J. B., Mahomoodally, F. M., Zengin, G., & Maggi, F. (2021). Essential oils as natural sources of fragrance compounds for cosmetics and cosmeceuticals. *Molecules*, *26*(3), 666.

Singh, A., Kapoor, R., & Misra, R. (2019). Green cosmetics-changing young consumer preference and reforming cosmetic industry. *International Journal of Recent Technology and Engineering (IJRTE)*, *8*, 12932.

Stone, H., Bleibaum, R. N., & Thomas, H. A. (2020). *Sensory evaluation practices*. Academic press.

Synthetic Biology in Everyday Life: From Lab to the Beauty Counter - SynBioBeta. (n.d.). Retrieved August 23, 2024, from https://www.synbiobeta.com/read/synthetic-biology-in-everyday-life-from-lab-to-the-beauty-counter

Wells, N. (2023). FDA Regulation of Cosmetics and Personal Care Products Under the Modernization of Cosmetics Regulation Act of 2022 (MoCRA). *Congressional Research Service (CRS) Reports and Issue Briefs*, NA–NA.

What are Chemical Exfoliants for Skin? | CeraVe Australia. (n.d.). Retrieved August 23, 2024, from https://www.cerave.com.au/blog/everyday-skin-care/what-are-chemical-exfoliants

What are essential oils? Uses and side effects. (n.d.). Retrieved August 29, 2024, from https://www.medicalnewstoday.com/articles/326732

Yadwade, R., Gharpure, S., & Ankamwar, B. (2021). Nanotechnology in cosmetics pros and cons. *Nano Express*, 2(2), 022003. https://doi.org/10.1088/2632-959X/ABF46B

Chapter 9

Trends in regulation and industry practices in essential oils and fragrances in cosmetics

Emerging cosmetic regulations and practice

Pallavi Shukla, Priya Verma, Vinita Tiwari, Anushka Tripathi, and Aradhana Mishra
Microbial Technology Division, CSIR- National Botanical Research Institute, Lucknow, India
and
Academy of Scientific and Innovative Research (AcSIR), Ghaziabad, India

Diksha Pathania
Academy of Scientific and Innovative Research (AcSIR), Ghaziabad, India
and
Photobiology Division, CSIR-Indian Institute of Toxicology Research, Lucknow, India

9.1 INTRODUCTION

Fragrance is a crucial component of cosmetics and is frequently considered by consumers to be the most significant factor when choosing a cosmetic. The odors of fatty acids, oils, and surfactants commonly employed in cosmetic formulations can likewise be effectively covered up by fragrances. In addition to adding pleasant smells to various products, essential oils (EOs) can function as active ingredients and preservatives while also providing several types of skin benefits, making them invaluable resources for the cosmetics sector. The regulation and industry practices surrounding EOs and fragrances in cosmetics have evolved significantly in response to consumer demand for natural ingredients and increased scrutiny regarding safety and efficacy. EOs are increasingly incorporated into cosmetic formulations due to their antimicrobial, anti-inflammatory, and antioxidant properties, which are beneficial for skin health and overall product appeal (1–3). The cosmetic industry is witnessing a shift toward eco-sustainability, with a growing emphasis on sourcing active ingredients from natural, renewable sources, as consumers

DOI: 10.1201/9781003558477-9

increasingly prefer products that are perceived as safer and more environmentally friendly (2). Despite the benefits, the use of EOs in cosmetics is not without challenges. The variability in the composition of EOs, influenced by factors such as plant genetics, growing conditions, and extraction methods, complicates their standardization and safety assessments (4, 5). This variability can lead to inconsistent product performance and potential adverse reactions, including allergic contact dermatitis, which has been documented among professionals frequently exposed to these substances, such as massage therapists (6, 7). Regulatory bodies are increasingly focused on ensuring that cosmetic products containing EOs are safe for consumer use, which necessitates rigorous testing and compliance with established safety standards (8). Moreover, the cosmetic industry is responding to regulatory pressures by adopting more stringent safety assessments for EOs. This includes evaluating the potential for sensitization and irritation, as well as the overall safety profile of fragrance mixtures that often contain EOs (8). The International Fragrance Association (IFRA) has implemented guidelines to limit the use of certain allergens and sensitizers in fragrance formulations, which directly impacts how EOs are utilized in cosmetic products (9). As a result, manufacturers are compelled to provide clear labeling and safety data to consumers, ensuring transparency and fostering trust in their products (4).

The cosmetic industry is shifting toward natural ingredients, with EOs and fragrances being regulated to ensure safety and efficacy, while evolving regulatory frameworks to promote safer and sustainable practices.

9.2 PREDICTIONS FOR FUTURE REGULATORY CHANGES

The regulation and industry practices surrounding EOs and fragrances in cosmetics are evolving rapidly, driven by increasing consumer demand for natural ingredients, heightened awareness of potential allergens, and stricter regulatory frameworks. The COVID-19 pandemic has further accelerated these trends, emphasizing the importance of safety and sustainability in cosmetic formulations (10). EOs serve not only as fragrance agents but also as preservatives and active components in cosmetic products. Their integration into formulations is largely motivated by consumer preferences for natural ingredients, which are perceived as safer alternatives to synthetic compounds (3, 11). The growing trend toward sustainability has prompted cosmetic companies to reformulate their products, prioritizing minimally processed natural materials to mitigate health concerns associated with synthetic fragrances (12). This shift is reflected in the increasing use of botanical extracts, which offer a vast array of bioactive compounds beneficial for skin health and overall product efficacy (5).

Regulatory frameworks are also adapting to these industry changes. In the European Union, for instance, the regulation of cosmetic products mandates the labeling of 26 specific fragrance allergens when present above certain concentrations, thereby enhancing consumer safety (13, 14). This regulatory scrutiny is crucial, as fragrances are among the most common allergens in cosmetics, with significant implications for public health (15, 16). The prevalence of allergic reactions to fragrances has prompted calls for more stringent regulations, including mandatory declarations of all fragrance ingredients and restrictions on the maximum allowable concentrations of known allergens (17).

Looking ahead, it is anticipated that regulatory changes will continue to evolve in response to consumer demands and scientific findings. The focus will likely shift towards enhancing transparency in ingredient labeling and improving safety assessments for cosmetic products to reduce the incidence of contact allergies (17, 18). Additionally, as the market for EOs expands, regulations may also address issues related to the adulteration of these oils, ensuring that consumers receive high-quality, authentic products (19).

The trends in regulation and industry practices concerning EOs and fragrances in cosmetics are characterized by a significant shift toward natural ingredients, enhanced safety measures, and increased transparency. Future regulatory changes will likely reflect these trends, aiming to protect consumers while fostering innovation in the cosmetic industry.

9.3 INDUSTRY TRENDS TOWARD SUSTAINABILITY AND TRANSPARENCY IN ESSENTIAL OILS AND FRAGRANCES IN COSMETICS

The cosmetics industry is undergoing a significant transformation toward sustainability and transparency, particularly in the realm of EOs and fragrances. This shift is largely driven by changing consumer preferences, heightened awareness of environmental issues, and the impact of the COVID-19 pandemic on consumer behavior. As consumers increasingly demand products that are not only effective but also environmentally responsible, cosmetic companies are compelled to adapt their sourcing, production, and marketing strategies to align with these expectations.

One of the most notable trends in the cosmetics industry is the growing preference for "clean beauty" products. This concept encompasses a range of practices aimed at minimizing carbon emissions, reducing water usage, recycling product containers, and decreasing waste throughout the product lifecycle. Research indicates that consumers are now more inclined to choose cosmetics that utilize sustainable ingredients and ethical sourcing practices, reflecting a broader societal shift toward environmental consciousness (10). The pandemic has further amplified this trend, as consumers have become more aware of the health implications of the ingredients in their cosmetics, leading to a preference for safer, natural alternatives (10, 20).

EOs, derived from various plants, have emerged as a popular choice for natural fragrances in cosmetics. The demand for these natural ingredients is driven by consumer concerns regarding the potential health risks associated with synthetic fragrance chemicals, which have been linked to various adverse effects (3, 21). EOs such as lavender, citrus, and eucalyptus not only provide appealing scents but also offer therapeutic benefits, making them attractive to both consumers and manufacturers alike (3, 5). As a result, many cosmetic companies are reformulating their products to incorporate these natural fragrances, thereby enhancing their market appeal while addressing consumer health concerns.

Moreover, the integration of sustainable marketing practices is becoming essential for cosmetic brands seeking to resonate with environmentally conscious consumers. Sustainable marketing encompasses strategies that prioritize environmental responsibility, ethical procurement, and transparent communication with consumers (22). This approach not only helps brands differentiate themselves in a crowded market but also fosters consumer trust and loyalty. The importance of transparency in ingredient sourcing and product formulation cannot be overstated, as consumers increasingly demand clarity regarding the origins and environmental impact of the products they purchase (23, 24).

The cosmetic industry's reliance on petroleum-based materials, particularly in packaging, has also come under scrutiny. The excessive use of plastics and microplastics poses significant environmental challenges, prompting a shift toward bio-based and biodegradable materials (25). Companies are now exploring innovative packaging solutions that align with circular economy principles, aiming to reduce their carbon footprint and promote sustainable consumption (26). This transition not only addresses environmental concerns but also enhances brand reputation among eco-conscious consumers.

In addition to packaging, the sourcing of raw materials is a critical area where sustainability efforts are being focused. The use of by-products from agro-industrial waste, such as cork from *Quercus suber*, exemplifies how the industry is leveraging sustainable practices to minimize waste and promote resource efficiency (26, 27). Furthermore, the ethical implications of ingredient sourcing are gaining attention, as consumers increasingly scrutinize the environmental and social impacts of their purchases (22). Brands that prioritize ethical sourcing and demonstrate accountability in their supply chains are likely to gain a competitive edge in the evolving market landscape.

The role of EOs in cosmetics extends beyond fragrance; they are also valued for their bioactive properties, which can enhance product efficacy (5, 28). As consumers seek multifunctional products that deliver both esthetic and health benefits, the demand for cosmetics containing essential oils is expected to rise. This trend is supported by research indicating that botanical extracts can serve as a rich source of active ingredients, offering a sustainable alternative to synthetic compounds (21, 28). Consequently, the cosmetics industry is witnessing a resurgence of interest in plant-derived ingredients, which aligns with the broader movement toward natural and organic products.

The regulatory landscape surrounding cosmetics is also evolving, with increased emphasis on safety and transparency. Regulatory bodies are advocating for clearer labeling of fragrance ingredients, including EOs, to ensure that consumers are informed about potential allergens and health risks (18). This push for transparency not only protects consumers but also encourages manufacturers to adopt safer formulations and practices. As a result, the industry is likely to see a shift toward more stringent regulations governing the use of fragrances and other cosmetic ingredients.

The trends toward sustainability and transparency in the cosmetics industry, particularly concerning EOs and fragrances, reflect a broader societal shift toward environmental responsibility and consumer health. As brands adapt to these changes, they must prioritize ethical sourcing, sustainable marketing practices, and transparency in their operations. The integration of natural ingredients, such as EOs, into cosmetic formulations not only meets consumer demand for safer products but also aligns with the industry's commitment to sustainability. Moving forward, the cosmetics industry will likely continue to evolve in response to these trends, shaping a more sustainable and transparent future for beauty products.

9.4 THE ROLE OF ADVOCACY AND CONSUMER EDUCATION IN ESSENTIAL OILS AND FRAGRANCES IN COSMETICS

The current consumer demands products with active ingredients extracted from natural sources that is influencing the cosmetic industry to explore vital new actives mainly from botanical extract. Since early times, EOs have been emerging as an effective natural ingredient in cosmetics, including fragrances, attributed to their wide array of individual component properties, e.g., antimicrobial, antioxidant as well as anti-inflammatory, thus the cosmetic industry is favoring the use of EOs as an active ingredient in various products ranging from skincare to perfumery (2).

The global demand of EO is facing several challenges, including:

1. low yield in respect to huge raw materials, due to which adulteration is raising significantly in recent years leading to safety issues (3),
2. standardization of dosage for minimizing the concern associated with the exposure to the EO ingredient (29), and
3. exhaustion of natural resources that may challenge the industry growth.

Advocacy groups play a critical role in addressing the challenges and concerns surrounding the use of EOs and fragrances in cosmetics by conducting ingredient research and educating consumers.

9.4.1 Advocacy groups and consumer education

The non-profit organizations like Environmental Working Group (EWG) advocates for safer personal care and cosmetic products including EOs and fragrances by maintaining a skin-deep database that reflects the number of scientific studies about the product or ingredient in the published scientific literature. Each of the products is evaluated based on the individual ingredient hazard, thereby promoting valuable transparency between product and consumer.

The Campaign for Safe Cosmetics (CSC) is a leading advocacy group that works to protect people and the environment from harmful chemicals in beauty products. It pursues this mission through four key strategies:

1. *Public education*: The CSC informs consumers that the term "fragrance" on a cosmetic label can encompass a wide range of synthetic and potentially harmful chemicals. It empowers consumers to read labels carefully and avoid products that simply list "fragrance" without disclosing the specific ingredients. The CSC advocates for cosmetics companies to fully disclose all fragrance components, rather than hiding them under the generic "fragrance" term.
2. *Highlighting fragrance*: The CSC cites research linking certain fragrance chemicals to issues like allergies, hormone disruption, and respiratory problems. It raises awareness about the potential health risks associated with fragrances, especially for vulnerable populations like those with sensitivities.
3. *Industry transformation*: As an early pioneer in the safe cosmetics movement, the CSC advocates for the cosmetics industry to phase out the use of dangerous chemicals linked to serious health issues, such as cancer.
4. *Legislative advocacy*: The CSC pushes for stricter regulations and updated laws that would mandate greater transparency and safety standards for cosmetic formulations.

Breast Cancer Prevention Partners (BCPP) is a non-profit organization that focuses on eliminating toxic chemicals, including those found in fragrances and EOs, from personal care and beauty products. BCPP reviews scientific research and raises awareness about the potential links between exposure to certain fragrance and EO chemicals and increased breast cancer risk. It specifically focuses on ingredients that may act as hormone disruptors or carcinogens.

9.5 IMPORTANCE OF CONSUMER AWARENESS AND EDUCATION

EOs are complex combinations of terpenes and phenylpropanoid chemicals found in many kinds of fragrant plants. These oils are often extracted by steam distillation, solvent extraction, or supercritical fluid extraction (30).

EOs have a wide range of applications, from fundamental constituents in fragrances and cosmetics to additives in animal feed, human meals, and beverages. Furthermore, these oils are used in many medical treatments due to their antibacterial, pain-relieving, relaxing, anti-inflammatory, antispasmodic, and local anesthetic qualities (31).

Although EOs are generally regarded as safe, as complex mixtures of compounds, some of them are known to be allergens and skin sensitizing agents; this needs to be indicated on cosmetics labels, especially for consumers with sensitive, allergic-prone skin or existing skin disorders, along with an advisory that they could opt for patch testing before using such products (3). Consumer understanding of EOs and fragrances in cosmetics is vital not only for making educated purchasing decisions, but also for assuring product safety, supporting ethical practices, and driving industry compliance with legislation. This chapter explores the role of consumer education in understanding the hazards and advantages of EOs and fragrances in cosmetics, as well as how this awareness influences regulatory and industry practices.

9.5.1 The growth of essential oils and fragrances in cosmetics

The use of EOs has grown dramatically in recent years, owing to customer desire for natural, unprocessed, organic, and synthetic-free products (32). This popularity was reflected in the expanding market for EOs, which was estimated at 247.08 kilotons in 2020 and is anticipated to rise at a compound annual growth rate of 7.5% from 2020 to 2027. Alongside the growing demand for EOs, there has been a surge of interest within the scientific community to investigate various facets of these oils. This interest encompasses multiple disciplines, including chemistry, biochemistry, biology, phytochemistry, physiology, and clinical research. Although there is still much to discover about EOs, recent scientific advancements have enhanced our understanding of their biosynthesis, active components, and their effects both *in vitro* and *in vivo*, as well as their application across different industries. The increasing interest in EOs, along with advancements in scientific research examining various aspects of these oils, has highlighted the need for regulation to ensure their safe and effective use. While legislative controls have not yet followed a fully structured and regulated framework, there have been efforts at both the national and international levels to regulate the use of EOs across sectors like healthcare, food, and cosmetics (33).

9.5.2 Regulatory frameworks governing essential oils and fragrances in cosmetics

Cosmetics are products designed to improve or maintain the appearance of the human body. They include skincare, haircare, perfumes, and personal hygiene products. While cosmetics might provide numerous benefits, they may also endanger consumers' health and safety. As a result, many countries have formed regulatory agencies to ensure the safety and efficacy of cosmetics. To safeguard consumers from potential danger, regulatory authorities set and enforce cosmetic product standards. They assess the safety and efficacy of cosmetic substances and develop recommendations for labeling, advertising, and packaging. Cosmetics are regulated by several main authorities, including the US Food and Drug Administration, the European Commission, and Health Canada. Regulatory organizations often require cosmetic producers to undergo safety tests. These evaluations include determining the potential toxicity, skin irritation, and other dangers related to the product or substance. Regulatory organizations also perform post-market monitoring to monitor adverse responses and ensure that products continue to meet safety and quality requirements. Furthermore, regulatory authorities play an important role in ensuring that cosmetics are safe and effective for users. The standards and recommendations they develop contribute to protecting customers from potential health concerns linked with cosmetic items. This chapter provides a quick overview of the key regulatory frameworks that govern cosmetic and herbal cosmetic products in various regions and countries. It is critical to follow these guidelines to ensure product safety and protect client interests. Companies must be vigilant as the cosmetics industry evolves and adapts (34).

9.5.3 Consumer awareness of essential oils and fragrance use

Considering the intricacies surrounding the use of EOs and fragrances in cosmetics, consumer education is critical for recognizing potential advantages and hazards. The following are the crucial areas where consumer understanding is required.

9.5.3.1 Toxicity and allergenicity

Photosensitization occurs when EOs containing phototoxins, such as furanocoumarins, are applied to the skin in the presence of sunlight or UVA light. These reactions can cause inflammatory skin reactions, including pigmentation, blistering, and severe burns. EOs should not be used on inflammatory or allergic skin conditions and should be diluted in vegetable oils. Exposure to the sun or tanning salon should be avoided for at least 24 hours after photosensitizing EOs (35, 36). Consumers must be educated on how to read labels, understand the potential risks associated with these substances, and select products that suit their skin types and sensitivities.

9.5.3.2 Adverse effects and contraindications

Side effects of EOs include irritation, photosensitization, and, in some cases, abortion (37). It's crucial to emphasize the need for education when using

EOs, especially during pregnancy, which is a vulnerable time. EOs, which are made up of small, liposoluble molecules, can enter the bloodstream after passing past the epidermal, pulmonary, and intestinal barriers, eventually reaching the placenta and accumulating (38) and even crossing the placental barrier (39). The placenta is a distinct, transitory organ shared by two organisms. It promotes appropriate fetal growth and development by coordinating nutrient and waste exchanges between the maternal and fetal circulatory systems, as well as allowing for bidirectional hormonal regulation of the mother and fetus (40). Exposure to hazardous molecules during pregnancy can cause placental malfunction, leading to illnesses including miscarriage or preeclampsia (41–44). This can have both short- and long-term effects on both the mother and the unborn child (45).

9.5.4 The role of consumer education in regulation compliance and industry practices

Consumer education plays a crucial role in ensuring compliance with regulations and shaping industry practices concerning EOs and fragrances. As the market for these products expands, driven by consumer interest in natural and health-related benefits, it becomes increasingly important to educate consumers about the proper use, safety, and regulatory standards associated with EOs. This education not only enhances consumer understanding but also fosters compliance with industry regulations, thereby promoting safer and more ethical practices. One of the primary areas where consumer education is vital is in understanding the therapeutic benefits and appropriate applications of EOs. For instance, a study has shown that specific formulations of EOs, such as those used in aromatherapy candles, can provide significant therapeutic benefits, including relaxation and stress relief, particularly at optimal concentrations (46). However, consumers must be educated about the correct concentrations and applications to ensure safety and efficacy. This is particularly important as the market is flooded with various products that may not meet safety standards or may be misrepresented in terms of their benefits (47). Educating consumers about these aspects can lead to more informed purchasing decisions and greater compliance with safety regulations. Moreover, the regulatory landscape for EOs is complex, particularly in regions like the European Union, where strict guidelines govern the production and marketing of these products (48). Consumer education initiatives can help demystify these regulations, enabling consumers to understand the significance of certifications and quality standards. For example, the acceptance of EOs as natural preservatives in food products hinges on consumer awareness of their safety and effectiveness compared to synthetic alternatives (49, 50). By informing consumers about the benefits and regulatory compliance of EOs, industry stakeholders can enhance consumer trust and promote adherence to safety standards. Additionally, the rise of misinformation, particularly during events like the COVID-19 pandemic, highlights the need for robust consumer education. Many consumers turned to EOs for purported health benefits against COVID-19, despite a lack of scientific evidence supporting these claims (51). This situation underscores the necessity for educational campaigns that provide accurate information regarding the efficacy and safety of EOs, thereby preventing misuse and promoting responsible consumption. Furthermore, consumer preferences significantly influence industry practices. Research indicates that factors such as brand image, product quality, and price play critical roles in consumers' purchasing decisions regarding EOs (52). By educating consumers about these factors, businesses can align their practices with consumer expectations, leading to improved product offerings and compliance with market demands. For instance, understanding the sensory acceptability of EOs in food products can guide manufacturers in developing formulations that meet consumer preferences while adhering to safety regulations (53, 54). In conclusion, consumer education is essential for promoting regulation compliance and improving industry practices related to EOs and fragrances. By enhancing consumer understanding of product safety, regulatory standards, and the implications of their choices, stakeholders can foster a more informed consumer base that supports ethical practices and compliance within the industry.

9.5.5 Challenges in consumer education for essential oils and fragrances

Consumer education regarding EOs and fragrances presents several challenges, primarily due to the complexity of these products, their potential for causing allergic reactions, and the misconceptions surrounding their safety and efficacy. While often perceived as natural and non-toxic alternatives to synthetic fragrances, EOs can pose significant health risks, particularly in sensitive individuals. One of the primary challenges in consumer education is the prevalence of allergic contact dermatitis associated with EOs. Studies have shown that EOs can contain numerous allergenic components, making it difficult for consumers to identify specific triggers (55, 56). For instance, individuals working in professions such as massage therapy are at a heightened risk of developing contact dermatitis due to frequent exposure to EOs (6, 7). Fragrance allergens are prevalent in many cosmetic products, and their exposure can lead to various health issues, including respiratory problems. For instance, fragrances can trigger asthma symptoms in susceptible individuals, as they often contain volatile organic compounds (VOCs) that can irritate the respiratory tract (13). The complexity of EO compositions, which can include numerous distinct constituents, complicates the identification of allergens and increases the likelihood of co-sensitization to multiple fragrances (55, 56). This underscores the need for consumers to be educated about the potential risks associated with using EOs, particularly for those with a history of allergies. Moreover, the perception that EOs are inherently safe can lead to misuse and overexposure. Many consumers may not be aware that

EOs can cause adverse reactions, especially when misused or in high concentrations (3). The misconception that "natural" equates to "safe" can result in consumers neglecting to perform patch tests or consult with healthcare professionals before using these products (57). This knowledge gap can lead to serious health implications, highlighting the necessity for comprehensive consumer education that addresses both the benefits and risks of EOs. Another significant challenge is the variability in the quality and composition of EOs available in the market. Factors such as extraction methods, plant source, and environmental conditions can significantly influence the chemical profile of EOs, which in turn affects their safety and efficacy. For example, the duration of steam distillation can alter the fragrance profile and potentially increase the concentration of undesirable compounds (58). Consumers often lack the knowledge to discern high-quality products from those that may be adulterated or improperly processed, which can lead to inconsistent experiences and health risks. Furthermore, the regulatory landscape surrounding EOs and fragrances is often less stringent than that for pharmaceuticals, leading to a proliferation of products with insufficient labeling or safety information (57, 59). This lack of regulation can result in consumers being misled about the safety and efficacy of EOs, further complicating their ability to make informed choices. In conclusion, the challenges in consumer education regarding EOs and fragrances are multifaceted, encompassing issues of allergen identification, misconceptions about safety, variability in product quality, and regulatory shortcomings. Addressing these challenges requires a concerted effort from manufacturers, healthcare professionals, and educators to provide clear, evidence-based information that empowers consumers to use EOs safely and effectively.

9.6 ROLE OF ADVOCACY GROUPS IN SHAPING REGULATIONS

Advocacy groups play a crucial role in shaping regulations concerning EOs and fragrances in cosmetics, primarily through their efforts to influence public policy and raise awareness about health and environmental concerns associated with these substances. Advocacy, at its core, involves the communication of specific viewpoints and the mobilization of resources to effect change in legislation and regulations (60). In the context of cosmetics, advocacy groups often focus on the safety and transparency of ingredient lists, particularly concerning potential allergens found in fragrance components. For instance, the European cosmetics regulation mandates that 26 allergenic fragrance substances must be disclosed on product labels if present above certain thresholds, which has been a point of advocacy for consumer safety groups (61).

The increasing demand for natural ingredients in cosmetics has also been a significant driver for advocacy groups. As consumers become more aware of the potential adverse health risks associated with synthetic fragrances, advocacy groups have pushed for the inclusion of EOs as safer alternatives in cosmetic formulations (3). This shift is not only a response to consumer preferences but also reflects a broader movement toward sustainability and natural products in the cosmetics industry (28). Advocacy efforts have led to greater scrutiny of synthetic chemicals, prompting companies to reformulate products to include more natural ingredients, thereby aligning with consumer demand and regulatory pressures (3, 28).

Furthermore, advocacy groups have been instrumental in highlighting the adverse health effects associated with synthetic fragrances, which can act as endocrine disruptors and cause allergic reactions in sensitive individuals (62, 63). This has led to calls for stricter regulations and the implementation of fragrance-free policies in public spaces, as many individuals report health issues when exposed to fragranced products (61). The advocacy for such policies underscores the importance of public health considerations in regulatory frameworks, emphasizing the need for accountability and transparency in the cosmetics industry. The role of advocacy groups extends to influencing regulatory bodies to adopt stricter standards for fragrance ingredients. For example, the IFRA has continuously amended its standards to lower recommended use concentrations for certain fragrance materials, responding to advocacy efforts aimed at reducing allergenic risks (6). Additionally, the European Union has implemented regulations prohibiting specific allergens in cosmetics, demonstrating the impact of advocacy on legislative changes (6, 63). Advocacy groups thus serve as a vital link between consumer concerns and regulatory action, ensuring that public health and safety remain at the forefront of cosmetic regulations.

9.7 CONCLUSION

The cosmetic industry is responding to consumer demand for natural, sustainable products by using plant-derived EOs, which offer both appealing scents and therapeutic benefits. Research suggests these natural oils can be effective alternatives to synthetic fragrances, which are linked to health risks. However, as companies reformulate with natural ingredients, they must address safety concerns, especially the risk of sensitization and allergic reactions. Regulations and industry practices around EOs and fragrances in cosmetics are increasingly focused on consumer safety and transparency. Growing concerns about allergic reactions to fragrance ingredients have prompted calls for implementing stricter labeling requirements. These regulations ensure consumers are aware of potential allergens. As regulations evolve, manufacturers must prioritize consumer safety and transparency, ensuring their products comply with standards and cater to the demand for natural and safe options. This focus on both regulation and industry practices will shape the future of cosmetics, emphasizing both efficacy and safety.

REFERENCES

1. Orchard A, van Vuuren S. Commercial essential oils as potential antimicrobials to treat skin diseases. Evid Based Complement Alternat Med. 2017;2017:4517971.
2. Guzmán E, Lucia A. Essential oils and their individual components in cosmetic products. Cosmetics. 2021;8:114.
3. Sharmeen JB, Mahomoodally FM, Zengin G, Maggi F. Essential oils as natural sources of fragrance compounds for cosmetics and cosmeceuticals. Molecules. 2021;26:666.
4. Mortimer S, Reeder M. Botanicals in dermatology: Essential oils, botanical allergens, and current regulatory practices. Dermatitis. 2016;27:317–324.
5. Sarkic A, Stappen I. Essential oils and their single compounds in cosmetics—A critical review. Cosmetics. 2018;5:11.
6. Brans R, Schröder-Kraft C, Bauer A, Weisshaar E, Skudlik C, Geier J, IVDK. Contact sensitizations in massage therapists with occupational contact dermatitis: Patch test data of the Information Network of Departments of Dermatology, 2008–2020. Contact Dermat. 2023;88:43–53.
7. Geier J, Schubert S, Reich K, Skudlik C, Ballmer-Weber B, Brehler R, Weisshaar E, Uter W, IVDK. Contact sensitization to essential oils: IVDK data of the years 2010–2019. Contact Dermat. 2022;87:71–80.
8. Dvořáková M, Svobodová L, Rucki M, Ševčík V, Hošíková B, Chrz J, Bendová H, Kejlová K, Očadlíková D, Malý M. The safety assessment of cosmetic perfumes by using in chemico and in vitro methods in combination with GC-MS/MS analysis. Altern to Lab Anim. 2023;51:224–248.
9. Calandra MJ, Wang Y. POV remediation agents: α-ketoglutarate salts and the treatment of citrus oils and citrus-based model fragrances. Flavour Fragr J. 2020;35:686–694.
10. Lee J, Kwon KH. Good ingredients from foods to vegan cosmetics after COVID-19 pandemic. J Cosmet Dermatol. 2022;21:3190–3199.
11. Sharma A, Kumar V, Mittal C, Rana V, Dabral K, Parveen G. Role of essential oil used pharmaceutical cosmetic product. J Res Appl Sci Biotechnol. 2023;2:147–157.
12. Fuloria S, Mehta J, Chandel A, Sekar M, Rani NNIM, Begum MY, Subramaniyan V, Chidambaram K, Thangavelu L, Nordin R. A comprehensive review on the therapeutic potential of *Curcuma longa* Linn. in relation to its major active constituent curcumin. Front Pharmacol. 2022;13:820806.
13. Lu C-H, Fang M-C, Chen Y-Z, Huang S-C, Wang D-Y. Quantitative analysis of fragrance allergens in various matrixes of cosmetics by liquid–liquid extraction and GC–MS. J Food Drug Anal. 2021;29:700.
14. Botvid S, Simonsen AB, Jørgensen C, Müller S, Kirkeby CS, Christensen AB, Johansen JD. Fragrance allergens in cosmetic products marketed for children in Denmark. Contact Dermatitis. 2023;89:374–381.
15. Liu Q, Zenobi R. Rapid analysis of fragrance allergens by dielectric barrier discharge ionization mass spectrometry. Rapid Commun Mass Spectrom. 2021;35:e9021.
16. Periyasamy MK, Sekar SC, Rai R. Analysis of hypersensitivity in fragrance series by patch testing. Indian Dermatol Online J. 2019;10:657–662.
17. Corsini E, Galbiati V. Contact allergy to fragrances: In vitro opportunities for safety assessment. Cosmetics. 2019;6:3.
18. Pastor-Nieto M-A, Gatica-Ortega M-E. Ubiquity, hazardous effects, and risk assessment of fragrances in consumer products. Curr Treat Options Allergy. 2021;8:21–41.

19. Urasaki Y, Le TT. Differentiation of essential oils using nanofluidic protein post-translational modification profiling. Molecules. 2019;24:2383.
20. Lee J, Kwon KH. The significant value of sustainable cosmetics fragrance in the spotlight after COVID-19. J Cosmet Dermatol. 2022;21:6540–6548.
21. Sharma A, Gumber K, Gohain A, Bhatia T, Sohal HS, Mutreja V, Bhardwaj G. Importance of essential oils and current trends in use of essential oils (aroma therapy, agrofood, and medicinal usage). Essent Oils. Elsevier; 2023. pp. 53–83.
22. Ajayi SA, Olaniyi OO, Oladoyinbo TO, Ajayi ND, Olaniyi FG. Sustainable sourcing of organic skincare ingredients: A critical analysis of ethical concerns and environmental implications. Asian J Adv Res Reports. 2024;18:65–91.
23. Lim HY, Kwon KH. Sustainable assessment of the environmental activities of major cosmetics and personal care companies. Sustainability. 2023;15:13286.
24. Guo M, Zhang L, He Q, Arabi SA, Zhao H, Chen W, Ye X, Liu D Synergistic antibacterial effects of ultrasound and thyme essential oils nanoemulsion against Escherichia coli O157:H7. Ultrason Sonochem. 2020;66:104988. doi: 10.1016/J.ULTSONCH.2020.104988.
25. Cubas ALV, Bianchet RT, Reis dos IMAS, Gouveia IC. Plastics and microplastic in the cosmetic industry: Aggregating sustainable actions aimed at alignment and interaction with UN sustainable development goals. Polymers (Basel). 2022;14:4576.
26. Dini I. “Edible beauty”: The evolution of environmentally Friendly cosmetics and packaging. Antioxidants. 2024;13:742.
27. Mota S, Pinto C, Cravo S, Rocha e Silva J, Afonso C, Sousa Lobo JM, Tiritan ME, Cidade H, Almeida IF. Quercus suber: A promising sustainable raw material for cosmetic application. Appl Sci. 2022;12:4604.
28. Dabral K. Role of essential oils and bioactive components for manufacturing cosmetic items. J Res Appl Sci Biotechnol. 2023;2:35–54.
29. Firenzuoli F, Jaitak V, Horvath G, Bassolé IHN, Setzer WN, Gori L. Essential oils: New perspectives in human health and wellness. Evidence-Based Complementary and Alternative Medicine eCAM. 2014;(2014):467363.
30. Ribeiro-Santos R, Andrade M, Sanches-Silva A, de Melo NR. Essential oils for food application: Natural substances with established biological activities. Food Bioprocess Technol. 2018;11:43–71.
31. Bakkali F, Averbeck S, Averbeck D, Idaomar M. Biological effects of essential oils–a review. Food Chem Toxicol. 2008;46:446–475.
32. Fonseca-Santos B, Corrêa MA, Chorilli M. Sustainability, natural and organic cosmetics: Consumer, products, efficacy, toxicological and regulatory considerations. Brazilian J Pharm Sci. 2015;51:17–26.
33. Naja F, Hamadeh R, Alameddine M. Regulatory frameworks for a safe and effective use of essential oils: A critical appraisal. Adv Biomed Heal Sci. 2022;1:7–12.
34. Mishra L, Kurmi BD. Cosmetics regulations and standardization guidelines. Pharmaspire. 2023;15:137–150.
35. Buckle J. Essential oils in healthcare. Elsevier; 2015.
36. Balick R. A peek into safe use of essential oils. Pharm Today. 2019;25:22–23.
37. Dosoky NS, Setzer WN. Maternal reproductive toxicity of some essential oils and their constituents. Int J Mol Sci. 2021;22:2380.
38. Gupta RC, Doss RB, Lall R, Srivastava A, Sinha A. Trichothecenes and zearalenone. Reprod Dev Toxicol. Elsevier; 2022;15(4):1003–1016.
39. Williams W. Pre-conception care and aromatherapy in pregnancy. J Clin Aromather. 2005;2:15–19.
40. Arévalo L, Campbell P. Placental effects on the maternal brain revealed by disrupted placental gene expression in mouse hybrids. Proc R Soc B. 2020;287:20192563.

41. Ball E, Bulmer JN, Ayis S, Lyall F, Robson SC. Late sporadic miscarriage is associated with abnormalities in spiral artery transformation and trophoblast invasion. J Pathol A J Pathol Soc Gt Britain Irel. 2006;208:535–542.
42. Burton GJ, Jauniaux E. Placental oxidative stress: From miscarriage to pre-eclampsia. J Soc Gynecol Investig. 2004;11:342–352.
43. Goldman-Wohl D, Yagel S. Regulation of trophoblast invasion: From normal implantation to pre-eclampsia. Mol Cell Endocrinol. 2002;187:233–238.
44. Sebire NJ, Foskett M, Fisher RA, Rees H, Seckl M, Newlands E. Risk of partial and complete hydatidiform molar pregnancy in relation to maternal age. BJOG An Int J Obstet Gynaecol. 2002;109:99–102.
45. Turbeville HR, Sasser JM. Preeclampsia beyond pregnancy: Long-term consequences for mother and child. Am J Physiol Physiol. 2020;318:F1315–F1326.
46. Kumalasari D, Husnayanti A, Pratiwi AP. Characteristic and taste test of aromatherapy candle from essential oil of Cananga odorata. Pharm Reports. 2023;3:52.
47. Rusdianto AS, Wiyono AE, Fiana FH. Aromatherapy oil massage formulation from essential oil: Tuberose flower (*Polianthes tuberosa*) and lime oil (*Citrus aurantifolia*). Int J Food, Agric Nat Resour. 2020;1:21–27.
48. El-Shemy H Potential of essential oils. BoD–Books on Demand; 2018.
49. Hanková K, Lupoměská P, Nový P, Všetečka D, Klouček P, Kouřimská L, Hlebová M, Božik M. Effect of conventional preservatives and essential oils on the survival and growth of Escherichia coli in vegetable sauces: A comparative study. Foods. 2023;12:2832.
50. Gökşen G, Gümüş P The future trend natural preservatives in the food system: Essential oils. Avrupa Bilim ve Teknol Derg. 2021;441–443.
51. Rahem A, Fadhilah AAN, Azwa F, Kusumastuti AR, Pradjarto CAN, Alvianto GH, Devi LPAT, Anjani NR, Prayoga RD, Rahmawati RIA. Pengaruh Edukasi Terhadap Pengetahuan Generasi Muda Di Media Sosial Terkait Peran Essential Oil Sebagai Upaya Preventif Covid-19. J Insa Farm Indones. 2022;5:43–55.
52. Goh JW, Ng AHH. Factors affecting online consumer buying behavior towards essential oils in Penang. Impact Glob Adv Technol Online Bus Model. IGI Global; 2021; 279–302.
53. Vital ACP, Guerrero A, Kempinski EMBC, de Oliveira Monteschio J, Sary C, Ramos TR, del Mar Campo M, do Prado IN. Consumer profile and acceptability of cooked beef steaks with edible and active coating containing oregano and rosemary essential oils. Meat Sci. 2018;143:153–158.
54. Celina A, Rahmawati D, Permana T. Application of lemongrass essential oil as a natural preservative agent for pineapple juice. PhD diss. 2018;2:69–78.
55. Ochi H, Kwong HL, Lim SPR. Leonine facies in a masseuse as a result of allergic contact dermatitis. Contact Dermat. 2021;84(4):268–270.
56. Corazza M, Amendolagine G, Borghi A, Toni G, Lauriola MM. Aromatherapy and occupational allergic contact dermatitis: Two further cases caused by lavender oil and other essential oils. Contact Dermat. 2019;81:378–379.
57. Irshad M, Subhani MA, Ali S, Hussain A. Essential oils-oils of nature. BoD—Books on Demand. 2020.
58. Petrović M, Petrović V, Mlinar Z, Babić S, Jukić J, Prebeg T, Kremer D Duration of steam distillation affects essential oil fractions in immortelle (*Helichrysum italicum*). Horticulturae. 2024;10:183.
59. Uter W, Schmidt E, Geier J, Lessmann H, Schnuch A, Frosch P. Contact allergy to essential oils: Current patch test results (2000–2008) from the Information Network of Departments of Dermatology (IVDK). Contact Dermat. 2010;63:277–283.

60. Miller EL, Roup CM. The importance of professional advocacy at the state level: A coalition model. Semin Hear. 2022;43(01):35–44.
61. Klaschka U. This perfume makes me sick, but I like it." Representative survey on health effects associated with fragrances. Environ Sci Eur. 2020;32:30.
62. Rádis-Baptista G. Do synthetic fragrances in personal care and household products impact indoor air quality and pose health risks? J Xenobiotics. 2023;13:121–131.
63. Bråred CJ, Hagvall L, Karlberg A-T. Fragrance allergens, overview with a focus on recent developments and understanding of abiotic and biotic activation. Cosmetics. 2016;3:19.

Chapter 10

Regulatory insights and recommendations for essential oils and fragrances in cosmetics

The key takeaways

A.B. Pant

CSIR-Indian Institute of Toxicology Research, Lucknow, India

10.1 AN OVERVIEW

This book investigates essential oils and fragrances in the cosmetics and personal care industries, covering their application, benefits, and regulatory requirements. It addresses various issues, including the historical use of essential oils, safety studies, and toxicological evaluations to ensure their safety in goods. The chapter also emphasizes the need for good manufacturing practices (GMP) to ensure product quality, safety, and consistency. It also delves into the complex regulatory landscape, providing practical advice on compliance solutions businesses may use to meet international skincare standards, labeling requirements, and advertising restrictions. By showcasing best practices for worldwide compliance, the book helps professionals navigate the regulatory environment more efficiently, ensuring that products fulfill both legal criteria and consumer expectations. The book focuses on market developments, consumer preferences, and the impact of new technology on the cosmetics sector. It explores how consumers are becoming more concerned with sustainability and transparency, and how businesses can modify their processes to suit these expectations. The significance of consumer education is also emphasized, since knowledgeable customers can better make decisions consistent with their beliefs and requirements.

Finally, the book is an invaluable reference for industry experts who want to stay current on the latest advancements of essential oils and fragrances. It provides practical recommendations for guaranteeing compliance, stimulating innovation, and fulfilling consumers' increasing requirements in a fast-changing industry. The book takes a progressive approach to negotiating the difficulties of the cosmetics industry, emphasizing sustainability, openness, and consumer education.

10.2 REGULATORY PRACTICES AND INDUSTRIAL TRENDS: A HISTORICAL PERSPECTIVE

Essential oils have a long and complicated history, stretching back centuries. Ancient civilizations used them for medical, spiritual, and cosmetic

DOI: 10.1201/9781003558477-10

purposes. Essential oils are now appreciated for their natural characteristics and are utilized in cosmetics, aromatherapy, and holistic treatments. Consumer desire for eco-friendly products has prompted the modern cosmetics business to focus on sustainability and natural ingredient sources. Essential oils with antibacterial and antifungal characteristics, such as orange, lemon, and citrus, are commonly utilized in skin and hair care products. Regulatory authorities and technological advancements have made essential oils safer and more efficient, resolving toxicity concerns while promoting their benefits. The future of crucial oil-based cosmetics is bright, with potential applications in developing practical and sustainable products that improve human health and wellness. This wonderful path has been shaped by changing regulatory policies, scientific breakthroughs, and consumer needs.

- *The early beginnings:* Essential oils have been utilized for ages in numerous civilizations for medical, spiritual, and cosmetic benefits. Ancient civilizations such as Egypt, Greece, Rome, and India used plant extracts in perfumery, cosmetics, and therapeutic applications.
- *The regulatory evolution:* As the cosmetics sector expanded, regulatory agencies arose. Governments developed standards and laws in the early twentieth century to ensure product safety and efficacy. Regulatory frameworks affecting the business include the United States Food and Drug Administration (FDA); the European Union's Cosmetics Directive; the Drugs and Cosmetics Act, 1940; and the Drugs and Cosmetics Rules, 1945 under India's Central Drugs Standard Control Organization (CDSCO).
- *The safety and toxicity:* Safety and toxicity were significant concerns in the mid-twentieth century. Regulatory bodies began to evaluate the potential dangers connected with scent compounds, creating guidelines and bans on specific substances.
- *Labeling and disclosure:* Labeling and disclosure have received considerable attention in recent decades. The European Union's Cosmetics Regulation compels manufacturers to declare fragrance allergens on product labels, encouraging transparency and customer safety.
- *Global harmonization:* The cosmetics sector has expanded globally, with companies operating in numerous locations. Regulatory harmonization efforts strive to harmonize regulations and facilitate international trade while protecting consumers.
- *Sustainability and natural ingredients:* The cosmetics industry is shifting toward sustainability and natural ingredients. Consumers demand environmentally friendly and responsibly sourced products, leading corporations to change their formulas and processes.
- *Future directions:* As the business evolves, regulatory processes will likely center on new challenges, including nanotechnology, endocrine disruptors, and environmental impact. Advocacy groups, scientific

studies, and consumer education will all play important roles in determining the future of essential oils and fragrances in cosmetics. Finally, the historical evolution of essential oils and fragrances in cosmetics illustrates a dynamic interplay of regulatory norms, scientific advancements, and customer desires. Understanding this history provides critical insights into the industry's complicated regulatory structure and ongoing efforts to balance innovation, safety, and sustainability.

10.3 RISKS AND BENEFITS OF ESSENTIAL OILS IN COSMETICS: A DELICATE BALANCE

Essential oils are popular in cosmetics because of their natural origin, attractive aroma, and potential health advantages. These oils provide various benefits, including anti-inflammatory and antibacterial qualities that help prevent dermatological problems, hair loss, and premature aging. Linalool, citronellol, alpha-bisabolol, eugenol, menthol, vanillin, citral, eucalyptol, limonene, and camphor are essential oils with several uses. Linalool, for example, has antibacterial, calming scent, and anti-inflammatory qualities. Citronellol contains antioxidant properties and protects the skin from UV rays. Alpha-bisabolol possesses depigmenting activities as well as antidepressant potential. However, essential oils have specific hazards, such as skin irritation and allergic reactions. Another area of concern is photosensitization. Quality, safety, and regulatory compliance are critical for reducing risks. Regulations such as REACH (Registration, Evaluation, Authorization and Restriction of Chemicals) and GMP guarantee safety and quality. Purity and composition are evaluated using modern analytical procedures such as HPLC, GC-MS/MS, LC-MS/MS, and FTIR. Safely using essential oils in cosmetics necessitates a comprehensive evaluation of potential dangers and benefits. The convincing data indicate that essential oils can be a possible alternative to chemical-based cosmetics regarding quality, safety, and regulatory compliance. Nanotechnology can improve cosmetic product efficiency and sustainability. Essential oils are significant in cosmetics because of their natural origin, possible advantages, and many qualities.

10.4 RISK ASSESSMENT AND MANAGEMENT OF FRAGRANCE ALLERGENS IN COSMETICS

Assessing and managing fragrance allergens in cosmetics is critical for consumer safety, as they may cause contact allergic dermatitis. The risk of skin sensitization was evaluated using a weight of evidence (WoE) methodology, which incorporated all relevant data into a three-tiered framework. Given the European Union's and many other nations' bans on animal testing for cosmetics, non-animal testing methods have emerged as the preferred method for toxicological assessment of scent compounds. The adverse outcome pathway (AOP) concept describes a sequence of events that lead to skin sensitization, and numerous *in vitro* and *in chemico* approaches have been validated to assess critical stages in this pathway. These procedures include the direct peptide reactivity assay (DPRA), KeratinoSens assay, human cell line activation test (h-CLAT), and the U937 cell line activation test. Acute toxicity testing, skin irritation testing using reconstructed human epidermis (RhE) models, phototoxicity testing, reproductive toxicity/endocrine disruption testing, mutagenicity/genotoxicity testing, and carcinogenicity testing using cell transformation assays (CTAs) are all used to test fragrances for safety.

The risk assessment of fragrance components is carried out in three steps. First, available data on the fragrance ingredients are thoroughly reviewed to determine their safety for relevant toxicological endpoints. If there are enough data, an endpoint evaluation is undertaken. If data are limited, the second phase employs a read-across approach in which the properties of related chemicals are analyzed to forecast the fragrance ingredient's safety. The third stage is to estimate the dermal sensitization threshold (DST), which defines a threshold below which the risk of sensitization is negligible. The no expected sensitization induction level (NESIL) is then utilized in a quantitative risk assessment (QRA) to determine the sensitizer's potency.

Allergen labeling is also required to limit the risk of exposure and prevent sensitization. The European Union has put legislation in place to label fragrance allergens in cosmetic items, and the Scientific Committee on Consumer Safety (SCCS) advises precise labeling criteria, such as concentration limits and individual labeling for specific allergies. Clear labeling, avoiding appealing packaging that may mislead consumers, and providing information on maximum concentrations are some risk management techniques. A smartphone software might potentially scan labels and rate their allergy severity. Finally, risk assessment and communication are critical for guaranteeing consumer safety, and recent advances in toxicological testing techniques and regulatory requirements allow for accurate hazard estimation and risk management. The cosmetics industry may ensure the safety of fragrance components and essential oils in cosmetic goods by using alternative testing methods and establishing appropriate risk management procedures.

10.5 THE ENVIRONMENTAL IMPACT OF FRAGRANCE COMPOUNDS IN COSMETICS

Fragrance compounds in cosmetics and personal care items may offer significant environmental risks due to their possible persistence, toxicity, and ecosystem bioaccumulation. These compounds are divided into natural and synthetic categories, with natural fragrances being more biodegradable than synthetic ones. However, even natural perfumes can have intricate chemical structures that decay slowly. Fragrance compounds' biodegradability is determined by their chemical structure, concentration, and environmental circumstances. Microorganisms play an essential role in biodegradation. However, some scent compounds can survive because of complex structures or antimicrobial characteristics. Synthetic scents, such as polycyclic musks (such as tonalide and galaxolide) and phthalates (such as dimethyl phthalate), pose special challenges because of their persistence and propensity for bioaccumulation and toxicity in aquatic and terrestrial ecosystems.

Regulatory frameworks such as REACH (European Union) and the Toxic Chemicals Control Act (TSCA) (United States) establish rules for testing and evaluating fragrance compounds' biodegradability and eco-toxicological impact. The International Fragrance Association (IFRA) also develops guidelines for testing scents and ensuring their safety in consumer products. Testing criteria and procedures, such as those provided by the Organization for Economic Cooperation and Development (OECD), are critical for determining scent compounds' biodegradability and eco-toxicity. The eco-toxicological impact of scent compounds is a significant problem, with many compounds threatening aquatic habitats due to their persistence, bioaccumulation, and toxicity. Synthetic musks and phthalates have been linked to endocrine disruption, reproductive and developmental difficulties, and other harmful impacts on aquatic creatures. Long-term research and legislative activities that promote biodegradable and environmentally friendly alternatives are necessary to solve these issues. Green chemistry-based solutions, such as generating non-toxic and biodegradable scent compounds, can help to address these concerns. Finally, the biodegradability and eco-toxicological impact of scent chemicals are serious concerns that regulatory agencies, industry, and researchers must address. By encouraging sustainable development and regulatory actions, we can limit the environmental impact of fragrance compounds and ensure a safer, more sustainable future. Advances in computer models and high-throughput screening approaches can help forecast eco-toxicological consequences and identify safer alternatives.

10.6 THE IMPORTANCE OF GOOD MANUFACTURING PRACTICES (GMP) IN COSMETICS

GMP are critical rules for guaranteeing the consistent production, safety, and quality of essential oils and fragrances in cosmetics. By following GMP, producers may ensure the highest quality standards, reduce contamination risks, and maintain regulatory compliance. GMP frameworks, such as ISO 22716 and regulatory requirements, specify essential features that producers must adhere to, such as people training, facility and equipment maintenance, raw material management, production process control, quality control, and documentation.

These rules ensure that items are consistently made and inspected to fulfill quality standards. GMP accreditation, gained through third-party audits, indicates conformity with established standards and ensures continuous quality and safety. This certification is critical for increasing consumer trust and competing globally. GMP compliance pushes the boundaries by allowing producers to differentiate themselves in a competitive market, gain consumer trust, and ensure safety and efficacy of their products. GMP certification benefits any manufacturer in the cosmetics sector that prioritizes product safety and quality. GMP compliance and certification are critical for cosmetics manufacturers that use essential oils and fragrances. It ensures that products are safe, effective, and of excellent quality, essential for gaining consumer trust and loyalty. By prioritizing GMP, firms may distinguish themselves as market leaders and achieve commercial success.

10.7 REGULATORY FRAMEWORKS AND INDUSTRY STANDARDS FOR ESSENTIAL OILS IN COSMETICS

The increasing demand for natural substances such as essential oils in the cosmetics business necessitates the development of creative solutions that balance advantages and potential hazards. Manufacturers must define essential oil quality parameters and follow global regulatory frameworks to assure quality and safety. The International Cooperation on Cosmetics Regulation (ICCR), OECD, and the International Organization for Standardization (ISO) collaborate to promote market growth, remove trade barriers, and foster innovation. These initiatives are intended to protect safety and quality of cosmetic items, particularly those that contain essential oils. Industry associations, such as IFRA, advocate safe use and regulatory compliance. IFRA members follow a code of practice that includes fragrance safety requirements, and the association makes recommendations for safe use to guarantee fragrance components and mixes are suitable for their intended use. Various organizations and groups govern the global cosmetic sector, ensuring product safety, quality, and regulatory compliance. Key regulatory authorities include the Association of South East Asian Nations (ASEAN), China's National Medical Products Administration (NMPA), Japan's Ministry of Health, Labor, and Welfare, and Health Canada. Certification organizations promote the standards for organic and natural cosmetics such as COSMOS, NATRUE, and Ecocert. At the same time, groups such as the Vegan Society and the Coalition for Consumer Information on Cosmetics (CCIC) provide vegan and cruelty-free product certification. These organizations collaborate to ensure the global safety, quality, and efficacy of cosmetic products and promote industry sustainability and transparency.

10.8 GLOBAL REGULATIONS ON ESSENTIAL OILS

Essential oils are regulated under various regulations and rules to ensure their safety in consumer products. The FDA's 21 CFR (Code of Federal Regulations) regulates food, medications, and cosmetics in the United States, mandating proper labeling and safety testing. The TSCA governs chemicals, including those in essential oils. The ISO sets essential oil standards, which include rules for characterization, residue evaluation, and water content determination. In the European Union, essential oils are subject to the REACH law, which compels enterprises to register their compounds and furnish safety data sheets (SDS). The European Union also has labeling and packaging restrictions, technical documentation requirements, and CITES (Convention on International Trade in Endangered Species of Wild Fauna and Flora) compliance.

Local authorities in China, Japan, and India that regulate essential oils include the NMPA, the Ministry of Health, Labour, and Welfare, and CDSCO, respectively. Companies must comply with these standards, which may require laboratory testing to confirm product safety and quality. To achieve compliance, businesses must stay current on local legislation, conduct rigorous safety assessments, apply GMP, provide correct labeling, and keep detailed records. Companies implementing these tactics can negotiate regulatory difficulties while ensuring the safety and quality of their essential oil-based goods worldwide. Sustainability policies are also crucial in the essential oil sector, with many companies promoting responsible sourcing, organic farming, and eco-friendly packaging. Companies combining regulatory compliance with sustainable practices can ensure long-term prosperity while conserving the environment.

10.9 INNOVATION AND SUSTAINABILITY IN ESSENTIAL OILS AND FRAGRANCES

Natural ingredients, notably essential oils and plant-based fragrances, are gaining popularity in the cosmetic business, driven by customer demand for mild, sustainable, and eco-friendly formulations. These natural substances may increase skin hydration, antibacterial capabilities, and heightened sensory experiences. The industry has investigated new technologies, including enhanced analytical procedures to verify purity and safety, and breakthroughs like nanotechnology, synthetic biology, and digital tools to improve the efficacy and stability of essential oils and fragrances. These advancements have altered cosmetic product composition, allowing for the speedy introduction of creative goods while ensuring compliance with safety laws.

Texture and aroma developments have revolutionized the sensory experience of beauty products. Companies are developing products that produce practical effects while providing a pleasant experience, focusing on multifunctional compounds and texture modifiers. Sustainability is increasingly important in influencing the future of essential oil and fragrance technology, focusing on environmentally responsible procedures, natural and renewable resources, and certification. Consumer impression is significant, given the growing need for transparently sourced and fairly traded essential oils. The industry must balance innovation, consumer protection, and environmental sustainability, promoting transparency, customer education, and thorough safety testing. By implementing these procedures, the industry can ensure that new technologies and ingredients are developed and adopted responsibly, and that these products are effective, safe, and ecologically friendly.

10.10 BALANCING INNOVATION, SAFETY, AND SUSTAINABILITY: THE FUTURE OF COSMETICS

Consumer demand for safer and more sustainable goods drives the cosmetics sector to embrace natural components, notably essential oils and perfumes. Essential oils are appealing alternatives to synthetic perfumes due to their therapeutic effects, which include antibacterial, antioxidant, and anti-inflammatory characteristics. Regulatory frameworks are expanding to safeguard consumer safety, including rules for fragrance allergen labeling and stricter cosmetic product guidelines. The industry is exploring novel extraction and formulation procedures to improve stability and efficacy while minimizing environmental impact. Consumer education is critical for promoting safe usage, explaining potential risks, and complying with rules. Advocacy groups are important for setting legislation, influencing public policy, and increasing awareness about the health and environmental risks associated with essential oils and fragrances. These organizations advocate for stricter regulations, transparency, and safety standards, eventually safeguarding consumers and encouraging industry accountability. The ability of the cosmetics business to combine innovation, safety, and sustainability while also addressing customer concerns and requirements will be critical to its future success. Prioritizing transparency, safety, and sustainability allows the industry to create effective, safe, and ecologically responsible products.

10.11 BEST PRACTICES FOR BUSINESSES NAVIGATING GLOBAL REGULATION

As cosmetics companies negotiate the complicated terrain of global regulations, implementing best practices is critical to maintaining compliance, product safety, and market success. Here are some major best practices that firms should consider:

1. *Stay informed and up-to-date*: Monitor regulatory updates, changes, and trends in different regions to guarantee compliance with current requirements.

2. *Global regulatory framework*: Create a worldwide regulatory framework with specific compliance criteria, standards, and processes for each region or country where products are sold.
3. *Product safety assessment*: Conduct a thorough product safety assessment to detect risks and ensure regulatory compliance.
4. *Good manufacturing practices (GMP)*: Follow GMP to assure product quality, safety, and consistency.
5. *Labeling and packaging compliance*: Ensure labeling and packaging meet regulatory requirements, including ingredient disclosure, allergen warnings, and language.
6. *Regulatory liaison*: Contact regulatory agencies, industry associations, and specialists to understand and influence regulatory decisions.
7. *Training and education*: Provide continual training and instruction for employees, distributors, and partners on regulatory compliance, product safety, and GMP standards.
8. *Documentation and record-keeping*: Maintain accurate and complete documentation and records for product development, testing, and compliance.
9. *Risk management*: Identify and address product formulation, production, and distribution hazards.
10. *Collaboration and industry engagement*: Collaborate with industry associations, regulatory authorities, and peers to share best practices, solve common difficulties, and influence regulatory decisions.
11. *Local regulations*: Understand local rules, cultural variances, and market requirements for compliance and product success.
12. *Supply chain management*: Ensure partners follow regulatory and GMP criteria.
13. *Product stewardship*: Manage product safety and environmental effects throughout its lifecycle.

10.12 BENEFITS OF BEST PRACTICES

Adopting best practices can help organizations comply with global standards and reduce the risk of non-compliance, improving product safety and quality. By proactively managing and adapting to regulatory changes, businesses can increase customer and regulatory trust, all while lowering compliance costs, improving efficiency, and maintaining competitiveness in a dynamic regulatory environment. Companies implementing these best practices may successfully traverse the complicated global regulatory landscape, assuring compliance, product safety, and market success. As the cosmetics industry evolves, it's apparent that the future of beauty depends on striking a careful balance between innovation, safety, and sustainability. Businesses can create goods that improve beauty and general well-being by stressing openness, consumer education, and environmental responsibility. As customers expect more from the companies they support, the industry must adapt to meet these demands, ushering in a new era of conscious beauty that helps both people and the environment.

Index